AF768692

CONTENTS

WITNESSES

TERMINATION OF THE SUPERCONDUCTING SUPER COLLIDER PROJECT

TUESDAY, MARCH 15, 1994

HOUSE OF REPRESENTATIVES,
COMMITTEE ON SCIENCE, SPACE, AND TECHNOLOGY,
SUBCOMMITTEE ON SCIENCE,
Washington, DC.

The subcommittee met, pursuant to recess, at 10:12 a.m. in room 2318, Rayburn House Office Building, Hon. Rick Boucher (chairman of the subcommittee) presiding.

Mr. BOUCHER. The subcommittee is now convened for the purpose of a hearing announced in pursuance of previous notice.

Six months ago the Congress voted to terminate the Superconducting Super Collider project. The Energy and Water Appropriations conference report for Fiscal Year 1994 instructed the Secretary of Energy to effect an orderly termination of the project and to submit to the Congress a plan to maximize the value of the investment that has been made in the project, including recommendations as to the feasibility of utilizing the Superconducting Super Collider assets, in whole or in part, in pursuance of an international high energy physics endeavor. That same conference report appropriated to the Department of Energy $640 million for Fiscal Year 1994 to carry out those instructions.

Today, the subcommittee will examine the three principal elements of the Department's termination process: the SSC termination plan prepared by the Department of Energy and submitted generally to the public and to this subcommittee, the Fiscal Year 1995 request by the Department of Energy for an additional $180 million for the SSC termination project, and an Agreement in Principle entered into between the Department and the State of Texas that was signed on March the 1st of 1994.

The Department's budget request for Fiscal Year 1995 includes, as indicated, a request for an additional $180 million for SSC termination. According to the SSC termination plan that was submitted by the Department last month, total termination costs for the project are expected to be $695 million. The Department currently has available a total of $735 million, which includes the appropriation last year of $640 million and a carryover from fiscal year 1993 of $95 million, and by its own estimate will have over $400 million available in previously appropriated funds for outlay in Fiscal Year 1995 to carry out the instructions contained in the conference report.

Therefore, the first question that this subcommittee will ask of today's witnesses is why an additional $180 million will be necessary in order to carry out the Department's obligations?

Secretary O'Leary has stated that the Department is committed to developing a follow-on project at the SSC site, and the Agreement in Principle creates a process to receive and evaluate proposals to establish a scientific research facility at that site. This is a generous and a good faith effort by the Secretary to minimize the loss suffered by the State of Texas as a result of SSC termination.

The subcommittee will ask, however, whether it is wise to make a commitment to a new research facility with estimated operating costs of between $20 million and $40 million on an annual basis at a time when the energy research budget will probably be level at best.

We will also ask whether the scientific community has recommended a new research facility in Texas, and we will ask whether scientists will be given an opportunity to review the scientific merit of that facility in comparison to the benefits of our other proposed research initiatives or to ongoing energy research programs.

Finally, in the Agreement in Principle executed between the Department of Energy and the State of Texas, the Department has estimated procedures to—or has established procedures to maximize the value of Superconducting Super Collider assets which appear to be limited to consideration of only those assets that are physically located at the Waxahachie site. We are interested in knowing what plans the Department has to maximize the value of off-site SSC-funded assets.

Many members of the scientific community believe that the results of the advanced detector and magnet research developed for the SSC at universities and at national laboratories will be an essential component of any future collaboration between the United States and European countries at the CERN facility and with respect to the Large Hadron Collider at CERN. Will these assets be preserved for later use, and what plans does the Department of Energy have with respect to those assets? We will look forward to the answers to these and other questions presented by our witnesses this morning.

And before turning to our first panel of witnesses, the chair would now recognize the gentleman from New York, Mr. Boehlert.

Mr. BOEHLERT. Thank you very much, Mr. Chairman.

I want to thank you for convening this hearing. I know you have not been eager to preside over the "shootout at the OK corral," but today's confrontation is necessary because the termination of the SSC raises stark issues of policy and principle, or perhaps I should say a lack of policy and a lack of principle, for the Department of Energy is handling the shutdown of the SSC with the same cavalier lack of concern for taxpayer interest that turned the Collider itself into such a disaster.

There must be some sort of miasma that pervades the corridors of the Forrestal Building, because no matter who is in charge the hallmarks of DOE involvement remain the same: disingenuous statements, questionable management, and obliviousness to or thwarting of the will of the House.

The will of the House stated in no uncertain terms during last year's debate was to shut down the SSC as quickly and as cheaply as possible, and to end the Federal presence at the SSC site. Here is what happened instead.

On February 7 the Department of Energy sent up two documents to Capitol Hill. One, the SSC termination plan, reported that DOE had more than enough money on hand to shut down the SSC. In fact, the agency would end Fiscal 1997 with a $40 million surplus in that account. And today we will hear that the surplus estimate is now $167 million.

The second document issued on February 7, the Department's Fiscal 1995 budget request, sought an additional $180 million for termination costs. Repeated efforts to ask DOE to explain how a surplus necessitates a need for additional spending have only produced embarrassed silence.

The agency is behaving like a little kid who was caught shoplifting even though he has money in his pocket. Worse still, DOE would like to institutionalize this sort of theft of tax money by committing to start yet another project, any other project, at the SSC site. This became clearer than ever 23 weeks ago when DOE suddenly signed an Agreement in Principle with Texas. No contract was ever so inaptly named. It ought to have been called "an agreement forsaking all principle."

Why are we rushing headlong into committing to build another project in Texas? Is it because some scientific need is going unmet? Is it because the Department has some policy goal in mind that would be furthered by use of this site? Is it because DOE has reviewed proposals and knows that one merits support? Is it because Texas signed on to the SSC because it was promised such a follow-up project? The answer to all of these questions is no.

The members of the High Energy Physics Advisory Panel, the group of scientists Texans and DOE officials used to cite with such authority just a few months ago, these distinguished scientists have expressed grave doubts about moving ahead with the new DOE lab in Texas, and well they might. Their existing programs are struggling for survival.

No, DOE is being propelled by neither scientific need nor legal obligation. The agency's motivation appears to be far more base. It is this simple. The Department of Energy seems to be more interested in Texas politics than in Federal interests.

Now, I know some will try to dismiss my comments as some sort of grudge match with Texas, or they will try to figure out some way New York could gain from SSC termination money. Or they may try to paint me as a partisan Republican despite my record and the fact that SSC issues have never split along party lines.

But I am motivated by the promise I made when I ended the fight against the SSC. I promised on the House floor to do all that I could to ensure that the demise of the SSC did not spell the doom of American high energy physics. I said that and I meant it.

Now, I don't blame Texans for what they are doing. They are fighting for their State's interest as any worthy State official or Member of Congress would do. I just wish that taxpayers had as dedicated, forceful and skilled an advocate in the Department of

Energy. That is the point the four citizen's groups that released statements this morning are making as well.

Spending money on payoffs to Texas will not help American science. Committing to a pure pork project for Texas will cripple American physics.

During the debate over the SSC the DOE and many Texans used to try to hide project mismanagement and parochial interest by cloaking themselves with the mantle of science. But today's hearing should demonstrate how thin a disguise that was. No one with the best interest of science or of the taxpayers at heart would be proceeding the way DOE is with termination of the SSC.

Thank you, Mr. Chairman.

Mr. BOUCHER. Thank you, Mr. Boehlert.

The gentlewoman from Texas, Ms. Johnson.

Ms. EDDIE BERNICE JOHNSON of Texas. Thank you, Mr. Chairman.

Mr. BOUCHER. All right. The gentleman from Texas, Mr. Barton?

Mr. BARTON. Thank you, Mr. Chairman. I am delighted that you are hosting this hearing. I think it is important and I know it is something that—it needs to be done.

I don't have any prepared remarks. I am going to speak from the heart and from the knowledge base that I have gained over the last 8 years in trying to be a small part of this project.

I don't think today's hearing is a shootout. I don't think it is a showdown. I think it is a necessary attempt to look at the status of the termination plan that the Department of Energy has prepared and sent to the Congress and to consider how that plan is being implemented.

I would point out that the SSC was never a Texas project. It was a national project with international significance. I would point out that we have a national laboratory in place in Texas that we have spent somewhere between $1.5 and $2 billion. I would point out that the State of Texas and its people by popular referendum voted in good faith to put up to $1 billion towards the construction and/or operation of that project, and I am told that the State of Texas has expended over $440 million in behalf of that contribution.

I don't think the Department of Energy has been derelict in its duty in trying to sell out to the State of Texas or anything like that. In fact, when it comes my time to ask questions I am going to mildly chastise the Department representatives for doing the opposite.

The 62-page report on termination had a half page on discussion of alternatives, and in the appropriation language that was passed last year the clear intent in voting not to build the SSC was to explore ways to maximize the assets, both people and tangible equipment, so that some benefit was obtained from the expenditure of the money that had been spent to date.

When the chairman of the Texas National Research Laboratory Commission has an opportunity to testify later in this hearing, that gentleman is not going to ask that we start any new projects. That gentleman is going to make some recommendations that have been reviewed out of about 60 to 70 that have been sent to the Governor's Advisory Committee on maximizing the assets, both people and equipment, that are already in place.

I don't view that as a new project start. I view that as we spent $1.5 to $2 billion. Let's get something out of it.

So, I would also point out to my good friend from New York that the vote last year in the Congress was not to eliminate a Federal presence in Texas. There was no discussion about that. The vote last year was to not build the SSC, and I abide by that vote, and every member of the Texas delegation and to my knowledge all other Members of Congress in the Senate and the House agree with that.

So, if my good friend from New York is looking for a shootout, it is going to be somebody armed against somebody unarmed, because I am not coming for a shootout. I want to hear the Department of Energy representatives. I want to hear the State of Texas representatives. I want an honest discussion of policy differences. And then this subcommittee and at some point in time the full committee, can make some decisions on how to proceed.

But this Member of Congress is not going to attempt to revive the Super Collider at this hearing or any other hearing in this Congress. This Member of Congress is going to attempt to make sure the money that has been spent has some positive use and that we don't do something counterproductive this year that forecloses the opportunity at some point in the future to revisit the issue of building the Super Collider.

At another hearing that this same subcommittee held earlier in the year, not one scientist, and we had both national and internationally renowned scientists, not one scientist said that there was not scientific merit in the Super Collider, nor did they say that any effort in Europe or anywhere else in the world would be a total and ultimate substitute for some day wanting to build a project like the Super Collider.

So, with that I will yield back to the chairman, and I look forward to the testimony and to the questions.

Mr. BOUCHER. Thank you very much, Mr. Barton.

The gentleman from Michigan.

Mr. SMITH of Michigan. Mr. Chairman, just a quick comment.

With our $1 to $2 billion investment, this is one of the larger projects in its beginning stages that is being terminated. The termination, I think, also is complicated because of the involvement and contribution of the State of Texas.

I am concerned about two areas: that we do this effectively in this termination, and that we be fair to Texas. That is my only comment, Mr. Chairman.

Mr. BOUCHER. Thank you, Mr. Smith.

[The prepared opening statement of Mr. Sam Johnson of Texas follows:]

OPENING STATEMENT—SUBCOMMITTEE ON SCIENCE, SAM JOHNSON, MARCH 15, 1994

Thank you Mr. Chairman, I would first like to welcome our distinguished panel here today and also thank the Chairman for holding these important hearings.

Four years ago the United States embarked on a historic mission that would revolutionize the science world. It would have made the United States a premier leader, uniting many of the brightest minds and the most sophisticated technology the science community has ever produced.

But after investing $2 billion, this project was abruptly halted by Congress. It was a bad decision for America. It is and will continue to be a loss for the science, medical and educational fields as well.

I hope now we can put aside our political bickering and continue to salvage what science is left from the cancellation of the SSC.

Three options have been generated by the Department of Energy that are viable and should be considered. We must not squander the entire $2 billion investment when facilities and equipment have been built and are operational.

The three options are: (1) to establish a regional medical technology center to conduct research for cancer. (2) to use the center for research in superconductivity. (3) and to operate a center for high-performance computing.

Let's proceed in a manner the reiterates the instructions that Congress gave to the Department of Energy—to maximize the value of the investment in the SSC while minimizing the loss to the United States.

The loss to the United States would be to stop funding the projects that can continue to benefit American people. So, I say to my colleagues, don't throw away tomorrow's science, let us work together to salvage some hope out of a project that held the dreams of a generation.

Thank you Mr. Chairman.

Mr. BOUCHER. The subcommittee is now pleased to welcome our first panel of witnesses today, led by the Honorable Martha Krebs, the Director of Energy Research at the U.S. Department of Energy. She is accompanied by Mr. James C. Hall, the Project Director for the Superconducting Super Collider Project Office, located in Waxahachie, Texas—that is a division of the Department of Energy—by Dr. John Peoples, the Director of the SSC Laboratory, and by the Honorable Robert Nordhaus, General Counsel of the Department of Energy.

Dr. Krebs, we welcome you this morning, and thank you once again for appearing before the subcommittee.

Without objection, your prepared written statement will be made a part of the record, and we would welcome your oral summary.

STATEMENT OF MARTHA A. KREBS, PH.D., DIRECTOR OF ENERGY RESEARCH, DEPARTMENT OF ENERGY, WASHINGTON, DC, ACCOMPANIED BY JAMES C. HALL, PROJECT DIRECTOR, DOE SSC PROJECT OFFICE, WAXAHACHIE, TX; JOHN PEOPLES, PH.D., DIRECTOR, SSC LABORATORY, WAXAHACHIE, TX; ROBERT R. NORDHAUS, GENERAL COUNSEL, DEPARTMENT OF ENERGY, WASHINGTON, DC; AND EDWARD CUMESTY, DEPUTY PROJECT MANAGER, SSC PROJECT OFFICE, WAXAHACHIE, TX

Dr. KREBS. Thank you. Mr. Chairman, members of the subcommittee, I would like to briefly summarize my testimony and introduce the discussion that I know is about to follow.

First of all, we have taken the termination requirements of the Congress with respect to the Superconducting Super Collider very seriously, and I am pleased to be here today with the team that the Department has put together in order to carry out the termination project.

First of all, Mr. Jim Hall from the Project Office, and Dr. John Peoples from URA have worked together in an exemplary fashion in carrying out the requirements of the termination project. So, in one sense, for us termination means carrying out things in a teamwork fashion.

It also means considering the people who are involved in the project and located at the site in Texas and other sites that were involved in the carrying out of the Superconducting Super Collider, and as a result the negotiation and development of the severance package was a very important element in taking care of the people involved in the SSC.

Finally, termination relies on accountability, and we believe that the termination plan submitted on February 7 demonstrates how we will carry out the termination in an accountable way with the appropriate cost and schedule controls.

I would like to make an addendum to my testimony. Towards the end of my testimony we note that the cost of the termination as indicated in the February 7th plan was $695 million. We also note that a validation process was ongoing. We have received the first reports of the results of that validation process. We have not looked at it very carefully at headquarters, but in the interest of complete openness we wanted to inform you that in fact the $695 million is likely to be reduced by $127 million to, I think, a total of $568 million.

The other aspect of termination required by the congressional decision was to maximize the benefits of the SSC assets. To that end we have carried out and agreed with Texas in the Agreement in Principle that was signed off by the Secretary and Mr. Shelton Smith from the Texas Collider Commission on March 1st. Within the Agreement in Principle we have agreed to establish a joint process to value and appraise the SSC assets, and this will be carried out by a panel that is mutually chosen.

We have agreed to good faith negotiations on the Texas claims. We have agreed on carrying out a process to develop expressions of interest which involves both a public request for expressions of interest and also working with Texas to assist their development of project definition studies and to guarantee them a fair evaluation in the light of the value of the assets and the future scientific directions for the country.

Within the Agreement in Principle we acknowledge that there is an opportunity to maximize value of the investment and to minimize loss to both Texas and the United States, and that the SSC assets and the site present—have present and future value that can be utilized in some ongoing manner that maximizes value for the United States and Texas.

In addition to those—to the Agreement in Principle, another aspect of our studying the way to maximize the benefits of the SSC assets for the Department and the United States is embodied in our appointment of the Drell Subpanel of the High Energy Physics Advisory Panel which is looking at the future vision for high energy physics in the long term beyond the next decade.

All of these efforts are intended to inform the July 1st report that is required, to inform Congress of how we plan to go about utilizing and maximizing the benefit of the assets at the SSC site.

Those conclude my remarks. However, I would like to introduce Mr. Ed Cumesty, Manager of the SSC Project Office in order for him to give you a very brief overview of the actual assets that are at the SSC sites, and these are—represent a considerable investment, and it is hard to encompass them without seeing some kind of representation of them. Mr. Cumesty will present them to you now.

Mr. BOUCHER. Thank you very much, Dr. Krebs.

Would you please state your name for the record?

Mr. CUMESTY. My name is Edward Cumesty, and I am the Deputy Project Director for DOE at the SSC site.

Mr. BOUCHER. We will be pleased to hear from you.

Mr. CUMESTY. Thank you. You may not be able to see, but you will hear from me.

I have some overhead transparencies.

Mr. BOUCHER. I wonder could you suspend just momentarily. I wonder if it might be possible for us to at least turn off the overhead lights so that we can have an opportunity to see these slides.

I think we pushed the wrong button. This is the Committee on Science and Technology, but like all others we occasionally make mistakes. There you are.

That is fine. Can you see?

All right. I wonder if we could ask some of our friends in the media if they would mind turning off the television light that is in that corner. Would that be to difficult to do?

I think that is probably good enough. That is all right. We can see.

Mr. CUMESTY. Is this microphone live then?

Mr. BOUCHER. I don't think it is.

Mr. CUMESTY. Well, I am live, so I will try to do without it.

This photograph is of the central facility. It is in Waxahachie, Texas, in Ellis County. It was purchased by the State of Texas and renovated by the State for use by the project.

It is essentially in three sections. At the far end is a warehouse complex. The building originally was a distribution center. And then the balance of the building is split pretty much 50/50 between a high bay and shop area and office area used by the staff of the laboratory.

The building is 557,000 square feet, and this is a typical picture on the office side of that facility. As you can see, it is set up with modular furniture and work stations. Approximately 1000 staff can be housed at this facility.

Mr. BOUCHER. Mr. Barton, we'll recognize you for questions in due course. Let's allow the gentleman to conclude his presentation, and you'll have your time.

Mr. BARTON. Mr. Chairman, that gives a false——

Mr. BOUCHER. Mr. Barton, you may point that out during the course of your questions, and we will recognize you for that purpose.

Yes, sir. You may proceed.

Mr. CUMESTY. Thank you, Mr. Chairman. I'll just hold this.

Moving from the central facility in Waxahachie out to an area known as the West Campus, and in particular the northwest corner of that piece of land at a site called N–15. This photo is about a year and a half old at this time. All of these facilities have been completed and they comprise a set of industrial activities that are key assets of the project.

I might point out the No. 6, which is the magnet development laboratory; No. 7, magnet test facility; and then in the center of the photo a series of large buildings in which is located the compressors and refrigeration equipment for the superconducting refrigeration system that would run the test facility in Building 7 and also provide cryogens, or would have provided cryogens to the Collider.

No. 5 is a shaft for magnet delivery, and if I might, there were in fact three shafts at 5, 9 and 8, and if you align those shafts, you get an idea of where the Collider tunnel was running under this site. The tunnel is completed. It is a lined tunnel in shale.

Moving to an interior shot of the Magnet Development Laboratory, it doesn't quite look like this. In fact, we have managed to straighten the walls out. But this facility was established to make specialty magnets and also to be a repair facility when the project was finally completed.

Magnets are assembled, starting in the lower right-hand corner and proceed forward in the building and make the turn, and these three magnets that you see setting here have been put in their cryostats. Then they would move out of the building in that direction. This building is completely facilitated and is fully operational.

If I can give you some perspective, the area that we were just looking at is the N–15 area which sets up here, and we are now looking north toward Dallas—or at least I am. It's a little hard to see. Sorry.

That's the N–15 area. Coming south into what is known as the injector area, what you see here at the center of the picture is the completed facility that would have housed the linear accelerator. It is a 2-story structure, and the bottom structure is where the machine would have been located.

And in this final picture, that is the linear accelerator enclosure, an 11 by 10 square tunnel which goes for about 800 feet, and is the initial—would have been the initial machine connecting them to the rest of the injectors and on into the Collider.

That concludes the photos I have got. Thank you.

Mr. BOUCHER. Thank you very much for the presentation.

Dr. Krebs, do you have anything else you would like to share with us by way of your opening statement this morning?

Dr. KREBS. No, sir. We are ready for questions.

[The prepared statement of Ms. Krebs follows:]

STATEMENT OF

MARTHA A. KREBS

DIRECTOR OF THE OFFICE OF ENERGY RESEARCH

DEPARTMENT OF ENERGY

BEFORE THE

COMMITTEE ON SCIENCE, SPACE, AND TECHNOLOGY

SUBCOMMITTEE ON SCIENCE

UNITED STATES HOUSE OF REPRESENTATIVES

MARCH 15, 1994

Mr. Chairman and Members of the Subcommittee:

Thank you for inviting me here today to discuss the termination of the Superconducting Super Collider. Six months ago, the Department of Energy, its prime contractor, Universities Research Association, Inc. (URA), and thousands of subcontractors nationwide were engaged in one of the largest scientific construction projects in history. With the official cancellation of the Super Collider on October 28, 1993, the Department commenced work on what now is one of the largest termination projects in history.

As you can well imagine, termination of this project is a massive effort, involving a myriad of tasks in areas such as future use assessments, personnel reductions, subcontract disposition, site restoration, and property disposition. While the Department regrets the decision to terminate the Super Collider, it respects and is bound by that decision, and is proceeding with the termination of this project in an efficient and effective manner.

The Department's development of a plan for termination of this project, as required by the Energy and Water Development Act of 1994 (Public Law 103-126), involves working with the State of Texas to maximize the value of the substantial investment made by the Federal government and the State of Texas in this project, while at the same time minimizing the losses to both the Federal government and the State of Texas from termination of the project. This plan involves three highly interrelated components: (1) termination of activities at the site and with associated subcontractors across the nation; (2) planning for future uses of the assets; and (3) working in good faith with the State of Texas to reach a prompt, fair, and reasonable settlement of the

Texas claim for reimbursement of its investment in the SSC so that the Department may proceed in a manner consistent with the Congressional mandate to orderly and expeditiously terminate this project. The framework for cooperation between the Department and the State of Texas in addressing these issues is contained in an Agreement in Principle which was signed on March 1, 1994, by Secretary O'Leary and Mr. Shelton Smith, Chairman of the Texas National Research Laboratory Commission.

This agreement represents a commitment by the Department of Energy and the State of Texas to make their best efforts at collaboration in resolving the complex legal and financial issues raised by the termination of this project. The State of Texas has identified more than $500 million worth of expenditures which it made in relation to the Super Collider, and has made a claim for reimbursement of these funds. The 1990 Memorandum of Agreement between the Department of Energy and the Texas National Research Laboratory Commission does not provide a clear path for resolving the Texas claim. It is the view of both parties that a constructive accommodation of interests is necessary in order to proceed with the orderly termination of this project, which is what the Agreement in Principle is intended to help facilitate. As appropriate, I will refer to the Agreement in Principle as I address the major elements of our termination activities below.

PLANNING AND EXECUTING "ORDERLY TERMINATION"

In the immediate weeks following termination of the Super Collider, a new team of senior managers was put in place to execute the termination phase of the project. A new project director, James Hall, was selected to manage the Department's activities at the Superconducting Super Collider Project Office (SSCPO). Mr. Hall brings extensive experience from his work for the Department as a senior manager at the Oak Ridge Operations Office, where he provided superb management in relation to development of the Uranium Enrichment Corporation and transition of the Department's enrichment activities to this new corporation. Leadership at the SSC Laboratory, managed by URA, also changed shortly after termination of the project. In early November, Dr. John Peoples, who serves as Director of the Fermi National Accelerator Laboratory, was named by URA to be the new Director of the SSC Laboratory. The Department is extremely pleased that both Mr. Hall and Dr. Peoples agreed to assume these responsibilities.

The Department also developed new roles and responsibilities for ensuring successful management of the Super Collider termination. As a first step, the SSC Project Office was put under the jurisdiction of the Oak Ridge Operations Office. This decision was made by Secretary O'Leary only after each of three candidate DOE field offices (the Oak Ridge Operations Office, Chicago Operations Office, and Albuquerque Operations Office) were thoroughly assessed for the strengths they would bring to the termination activities of this project. The Oak Ridge Operations Office was selected primarily due to its successful management of several previous large termination projects, including termination in the early 1980s of the Clinch River Breeder Reactor.

The new delineation of roles and responsibilities for the project puts the SSC Project Office in charge of the day-to-day mechanics of project termination,

80-521 O - 94 - 2

with technical and administrative back-up support provided by the Oak Ridge Operations Office, and with oversight at Headquarters from the Office of the Associate Deputy Secretary for Field Management. The Office of Energy Research has responsibility for issues associated with maximizing the value of the assets of the SSC project and for planning for the future of high energy physics research now that the SSC has been terminated. The Office of General Counsel is responsible for legal issues, while overall policy direction is provided by the Office of the Secretary.

With these changes in personnel and organizational management, we believe that the Department and the SSC Laboratory have the experienced leadership and administrative support necessary to successfully address the full range of activities required to achieve "orderly termination," as required by Public Law 103-126, the Energy and Water Development Appropriations Act, 1994.

The laboratory has already undergone a rapid transition from design and construction to project termination. The Department notified URA of the termination of its contract on November 17, 1993. Under its contract with the Department, URA is required to terminate all its commitments. The November 17, 1993, letter to URA also directed the SSC Laboratory to prepare a detailed task-oriented termination plan with costs, workforce requirements and schedules, based on a new termination budget structure that was provided in the letter. URA's termination plan was developed during December and January, and was transmitted to Congress on February 7, 1994, under the joint signature of James Hall and John Peoples. This plan only addresses costs and resource requirements associated with termination activities, and does not include any estimates associated with future uses of the Super Collider assets, follow-on projects, or the Department's efforts to reach a prompt, fair, and reasonable settlement of the Texas claim for reimbursement of its investment in the Super Collider. These latter elements require resolution in association with the Department's termination activities but cannot at present be estimated.

The executive summary of the February 7, 1994, termination plan states in the first sentence that the document was the result of a cooperative effort between the Department of Energy and Universities Research Association, Inc. I am pleased to tell the Subcommittee that cooperation between the Department and URA throughout the termination activities to date has been excellent, representing the type of working relationship that will be essential for the purpose of terminating the project in an orderly fashion.

The February 7, 1994 termination plan presents the Work Breakdown Structure (WBS) that will be used to control and manage the budget and tasks required during project termination. The eight major WBS categories are:

WBS 1: Administration and support captures the activities and costs associated with administrative and support functions of the SSC Laboratory which include planning activities, cost and schedule systems, and personnel administration.

WBS 2: Personnel demobilization captures the costs associated with separation of the workforce from the SSC Laboratory and designated subcontractors, both in terms of termination assistance and, for on-site employees, outplacement assistance.

WBS 3: <u>Documentation and technical close out</u> captures the costs of close out of the scientific and technical activities of the SSC Laboratory. This includes tasks associated with the compilation, editing, and indexing of a comprehensive reference document which will summarize the status of the work that was undertaken on the research, development, design, and construction of the SSC as of the time of termination. This WBS element also captures costs for completion of in-progress scientific and technical work which the Department believes is important for maximizing the value of the work done to date.

WBS 4: <u>Contract close out</u> captures the expenses for orderly termination and close out of the commitments such as subcontracts, Memorandum Purchase Orders (MPOs), and agreements at non-U.S. laboratories.

WBS 5: <u>Asset phase out</u> captures the costs for disposition of physical and other property acquired by the SSC Laboratory, as well as site restoration activities. Assets include real and personal property such as facilities, land, capital equipment, administratively-controlled property, and intellectual property, including software. Implementation of an asset phase-out plan will be addressed cooperatively between the Department of Energy and the State of Texas. The Department will develop a Site Restoration Plan, which will comply with the regulatory requirements of federal, state, and local agencies having jurisdiction. Long-term restoration activities will depend on the ultimate disposition of the SSC facilities.

WBS 6: <u>Environment safety and health program</u> captures the costs associated with environmental, safety, and health related planning, documentation, and compliance activities, including data collection and monitoring, inspections, permit application, and occupational medical services that may be required during termination of the project.

WBS 7: <u>Economic development and facilities use</u> captures costs associated with studies and evaluation of options for the use and disposition of the scientific and technical assets and facilities of the SSC Laboratory, particularly in relation to completion of the report required by law to be submitted by July 1, 1994, from the Secretary to Congress on future uses of the scientific and technical assets of the SSC.

WBS 8: <u>Facilities maintenance and security</u> captures the costs of ensuring proper maintenance and security of all occupied and non-occupied SSC Laboratory facilities and property. With many facilities spread over 200 square miles, and with well over 100,000 pieces of controlled personal property the requirement for this WBS element is to protect property against damage, theft, or destruction until the final disposition of the property has been determined.

PERSONNEL REDUCTIONS AND TERMINATION ASSISTANCE

One of the first challenges which the Department faced in terminating this project was to develop a termination assistance package for SSC employees. The specific directive from Congress to the Secretary was to provide termination assistance involving up to 90 days termination pay and reasonable relocation assistance to full-time employees of designated contractors and subcontractors. In response, the Secretary designated regular, full-time employees of Universities Research Association and its integrated on-site subcontractors, as well as the full-time employees at the three major off-site magnet development facilities, as eligible recipients of termination assistance. This assistance is comprised of 60-days severance pay, a $15,000 dislocation allowance, and limited extension of medical, dental, and life insurance. The Department developed this assistance package after extensive interaction with affected employees. We believe the package is responsive to the needs of the employees while also being responsible to the American taxpayers.

The termination of the SSC is forcing the Department to close its newest laboratory and terminate over 2,000 employees. This is a precipitous and unprecedented action, not only for the Department, but for the nation as a whole. Never before has the United States initiated a major new scientific laboratory, recruited more than 2,000 employees to operate the laboratory, and then terminated the laboratory--all over the course of less than five years. The great majority of the people who are losing their jobs at the Super Collider were hired since 1991. A large number of these employees left secure positions in academia, industry, and other Department of Energy laboratories. Their career decisions to work on the SSC were based on the excitement of contributing to this national initiative and expectations of enduring employment. Since the project would not have been completed until approximately the year 2000, and would have had an operational life of 20 years or more, employees who moved to Waxahachie had every expectation of remaining there for perhaps several decades.

Approximately 60 percent of the SSC Laboratory employees uprooted themselves and their families to take jobs at the SSC. Many of these employees, through no fault of their own, will suffer serious financial hardship. For the majority of them, reemployment in the Waxahachie area is unlikely, thus they will have to relocate to find new jobs. Some will have to sell homes at a significant loss. The dislocation allowance will help dampen the financial impact on many of these households.

Put in context, the SSC termination assistance package is within the range of assistance offered during other workforce reductions at DOE. The severance benefits for the Super Collider are projected to cost an average of $29,000 per employee. In comparison, at our Kansas City Production Facility and Oak Ridge Laboratory, both of which are experiencing reductions-in-force, the cost per separation is running at approximately $33,000. At Rocky Flats and Savannah River, the cost per separation has been $23,000 and $25,000, respectively. The major difference is that, rather than setting specific amounts for retraining and relocation (normally up to $10,000 per employee for retraining and $5,000 for relocation), the Department is providing the lump sum dislocation allowance of $15,000 per employee. This assistance addresses the primary need of flexible assistance that was expressed to the Department during lengthy meetings with employees. The flexibility provided in this termination assistance package allows the workers to determine how best to meet their own needs (retraining, relocation, home equity loss, professional job placement services, etc.).

TERMINATION SCHEDULE AND BUDGET

The termination plan projects that essentially all URA termination activities will be completed by October 1995, with any remaining URA responsibilities for settlement of subcontract termination claims to be executed by a small number of personnel who will remain employed by URA but paid by and located at Fermi National Accelerator Laboratory, which, like the SSC Laboratory, is also managed and operated by URA. After October 1995, the Department will assume responsibility for any remaining on-site activities. Site restoration activities, which will be managed by the Department, are expected to continue until at least September 30, 1996, and perhaps longer, depending on the results of the environmental analysis currently being conducted.

The termination plan is supported by detailed budget and scheduling estimates. These detailed estimates are currently being validated by DOE to establish a refined baseline for project termination. The estimated cost of the tasks listed in the plan totals approximately $559 million. To this subtotal must be added the costs incurred in October and November 1993 ($51 million), and an allowance for DOE management (Program Direction, $30 million) and for DOE-controlled contingency (approximately $55 million), for a total estimate of $695 million through Fiscal Year (FY) 1997.

As mentioned earlier, the termination plan budget does not include the full scope of activities that will be necessary to maximize the value of the substantial investment made by the federal government in this project while at the same time minimizing the losses to both the federal government and the State of Texas resulting from termination of the project. As stated in the Agreement of Principle, the Department and the State of Texas are working together both to define future uses of the assets and to reach a prompt, fair, and reasonable settlement of the Texas claim for reimbursement of its investment in the Super Collider.

PLANNING FUTURE USES OF THE ASSETS

Let me turn now to the process of examining future uses of the assets of the SSC. As you know, the Energy and Water Development Act of 1994 directs the Secretary of Energy to prepare and submit a report to Congress by July 1, 1994, with recommendations on how best to "maximize the value of the investment" that has been made in the Super Collider, including recommendations as to "the feasibility of utilizing SSC assets in whole or in part in pursuit of an international high energy physics endeavor."

Before discussing the specific approach which the Department sees for examining future uses of the SSC assets, let me first provide you with a snapshot of where the project was technically at the time of its termination.

Although for many people the Super Collider was a distant abstraction, for the thousands of people working on this project in practically every state of the nation, it was a massive scientific construction project that was making substantial progress when it was terminated last October. The nearly $2 billion in Federal and Texas funding had brought the project 20 percent toward completion, with considerable progress to show for that investment. Examples of that progress include:

- Acquisition and complete refurbishment of a 500,000 square foot Central Facility in Waxahachie, Texas, which contains modern office space for 1,000 employees; dozens of fully equipped electronics, computation, manufacturing, and design laboratories; as well as a 100,000 square-foot warehouse area.

- Successful execution in August 1992, of a test, under full operating conditions, of a 100-meter long string of superconducting magnets at the Accelerator Systems String Test facility. The test was accomplished six weeks ahead of schedule and represented a major advance in U.S. superconducting magnet design.

- Excavation of over 14 miles of the main ring tunnel, which involved setting new world records for number of feet of tunnel excavated in one day.

- Significant magnet design and production activities throughout the project. Collider dipole magnets and quadrupole magnets had passed Preliminary Design Reviews. Prototype collider dipole magnets and quadrupole magnets had been built and tested. The testing had confirmed the collider magnet designs, and manufacturing had begun of pre-production model collider dipole and quadrupole magnets. This superconducting magnet technology had been transferred to industry and large facilities for industrially manufacturing these magnets were being proofed with prototype magnets.

- Construction of a highly sophisticated Magnet Development Laboratory on site, with large cryogenics facilities necessary for superconducting magnet operation.

- Establishment of a computer network which supported detector design, accelerator simulation, and magnet design activities. Computational resources at the laboratory include the Physics Detector Simulation Facility, which can provide parallel computing of 7,000 million instructions per second, as well as some 3,000 personal computers and work stations.

- Acquisition by the State of Texas, through the Texas National Research Laboratory Commission, of 16,000 acres of land necessary for the 54-mile SSC accelerator.

- Finally, staff of over 2,000 technical and administrative personnel had been assembled from around the world, many of them leaders in their respective fields.

Although the project was terminated, the Department of Energy firmly believes that the scientific and technical resources developed for the Super Collider represent a major resource that should be devoted to future uses. The Department already has received more than 80 written suggestions on how to make use of the SSC assets. Both the Department of Energy and the State of Texas have an interest in maximizing the value of the investment that has been made in the Super Collider. Until recently, however, uncertainties regarding ownership of the assets have delayed development of a strategy for productive utilization of these assets. However, the March 1, 1994, Agreement in Principle between the Department and the State of Texas establishes a framework for moving forward on this issue.

Specifically, the Department and the Texas National Research Laboratory Commission (TNRLC) have agreed that the Commission will submit to the Department several recommendations for candidate uses of on-site assets, and the Department will fund the development of detailed feasibility studies for these candidate uses. Specific recommendations in three areas have been submitted to Governor Ann Richards jointly by the TNRLC and the Governor's

Advisory Committee, headed by Mr Jess Hay, which was established by the Governor in December for the purpose of examining options for future uses of the SSC assets. During the next two weeks, the Department and the TNRLC will finalize a specific strategy for dedicating funds toward the development of project feasibility studies for the three areas jointly recommended by the Jess Hay advisory committee and the TNRLC. These studies will include budgetary requirements, implementation schedules, workforce requirements, technical background, business plans, and other information that will be necessary for providing a full evaluation of each candidate use.

The Department and the Commission have also jointly issued a public call for Expressions of Interest (EOI) in the on-site use of SSC assets to generate additional suggestions of how the assets might be utilized. The call for EOI's has been published in the Commerce Business Daily, and sets an April 15, 1994 deadline for submissions. After the Expressions of Interest are received, panels of experts will be assembled from appropriate disciplines to review them. The Department will provide some funding in the form of grants for project definition studies of attractive concepts identified during these reviews. Independent peer review panels will then evaluate all of the proposals prepared by the Commission and the grantees.

Another component of the Department's planning for future uses of assets associated with the Super Collider involves the High Energy Physics Advisory Panel (HEPAP), which, as I mentioned during my testimony before you on January 26, 1994, has been asked by the Secretary to define "a long term program to pursue the most important high energy physics goals now that the

SSC has been terminated." HEPAP has created a subpanel for this purpose, chaired by Professor Sidney Drell, and among the areas being considered by this group is the Large Hadron Collider (LHC). The LHC, planned for construction at the CERN laboratory in Geneva, Switzerland, is the most promising large international high energy physics project in which the United States could cooperate in the near term. The LHC will consist of a new ring of magnets installed in the existing LEP tunnel, and will make available proton-proton collisions at 7 trillion electron volts (TeV) on 7 TeV. Although well below the energy of the SSC, (20 TeV on 20 TeV), it will, nevertheless, be the most powerful accelerator in the world by a substantial margin. Construction is expected to begin in 1995 and be completed in 2002.

HEPAP's recommendation regarding the LHC--or any other international collaboration--in conjunction with responses to our call for Expressions of Interest, and proposals by the Texas National Research Laboratory Commission, will serve as the basis for our recommendations regarding the use of SSC assets in an international collaborative venture or for other purposes.

The recommendations and plan which the Secretary will submit to Congress this summer will be aimed at recovering benefits from the investments that were made in the project. These recommendations, however, will be of little use unless acted upon quickly. Without funds appropriated for the purpose, it will not be possible to implement any proposals until FY 1996. If funding were delayed until then, and we continued our existing termination schedule, many future use options would be jeopardized by the erosion of assets. Human assets are perhaps the most seriously at risk.

The scientific, engineering, and computer programming workforce has dropped from a level of approximately 1,000 in October 1993 to less than 350 today, and currently is projected to drop to only 80 as of July 1, 1994. Such a steep drop, and further reductions currently planned for later this year, could seriously jeopardize some options for future use of the Super Collider assets. As a result, the Department is examining options for modifying the existing termination plan to maintain a somewhat larger staff of scientists, engineers, and other technical workers at the site, so that there exists a cadre of technical staff who could help implement recommendations submitted by the Department to Congress in our July 1994 report. Although we have not yet made any decisions regarding the specific number, duration, or cost associated with such an action, we will keep Congress informed when any decisions are made in this regard.

Regarding funding for possible future uses of the Super Collider assets, this consideration was included in development of the President's budget request of $180 million for FY 1995. The Department currently has $735 million in available funds for termination of the Super Collider ($640 million in FY 1994 appropriations, $65 million in termination liability appropriated in FY 1993, and $30 million in other commitments). Against this funding, the Department currently has $695 million in expected costs through FY 1997, excluding the expenses that will be required for reaching a reasonable settlement of the claim by the State of Texas and for future uses of the Super Collider assets. The FY 1995 funding request of $180 million will be necessary for addressing these additional termination expenses. Final estimates for these expenses have not yet been determined, and are subject to the ongoing discussions between the Department and the State of Texas. However, as additional details become apparent, the Department will directly inform the Congress.

In closing, Mr. Chairman, I stress the Department's commitment to terminate the Super Collider Project in a cost effective manner, and my own desire to work cooperatively with you and the other Members of this Subcommittee in both the termination of the SSC and the development of a new vision for the high energy physics program in the wake of the SSC termination decision.

This concludes my prepared statement. I would be happy to answer any questions you may have.

Superconducting Super Collider Project
N-15 Construction Site

1. ASST Compressor Building
2. ASST String Enclosure
3. ASST Cryogenic Refrigeration Service Building
4. Magnet Power Supply Control Room
5. Magnet Delivery Shaft
6. Magnet Development Laboratory
7. Magnet Test Laboratory
8. Personnel Shaft
9. Utility Shaft

80-521 O - 94 - 3

VMS 92-6-351-11
7-8-92
MDL Interior [redacted]

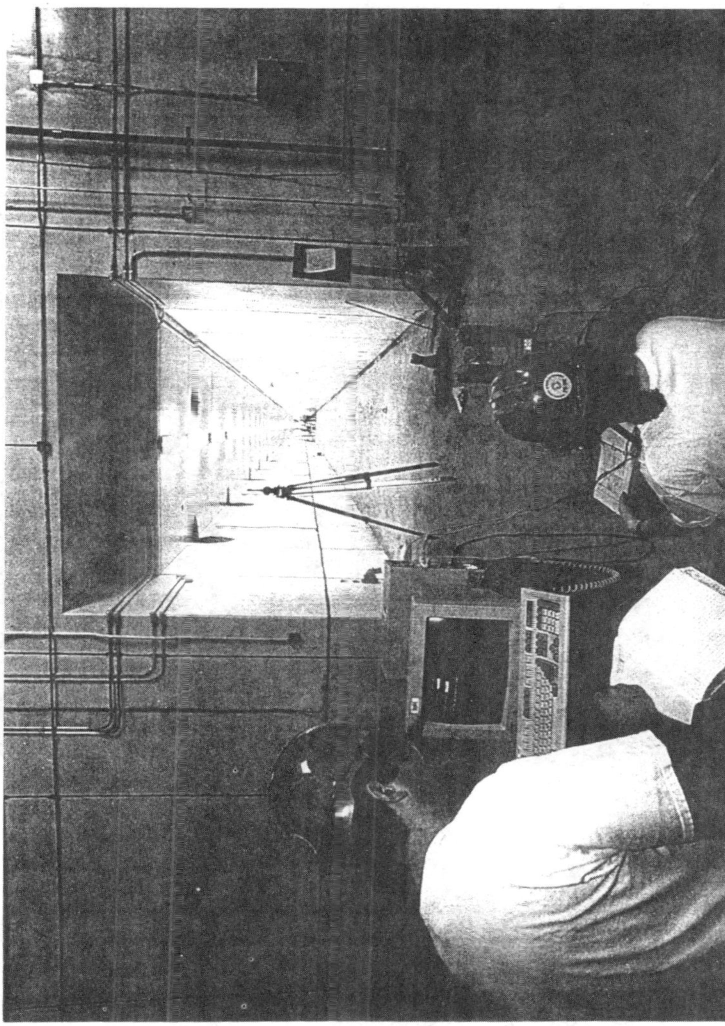

VMS 93-4-58-4
5-19-93
LINAC

92-6-272-4
9-24-92
Central Facility

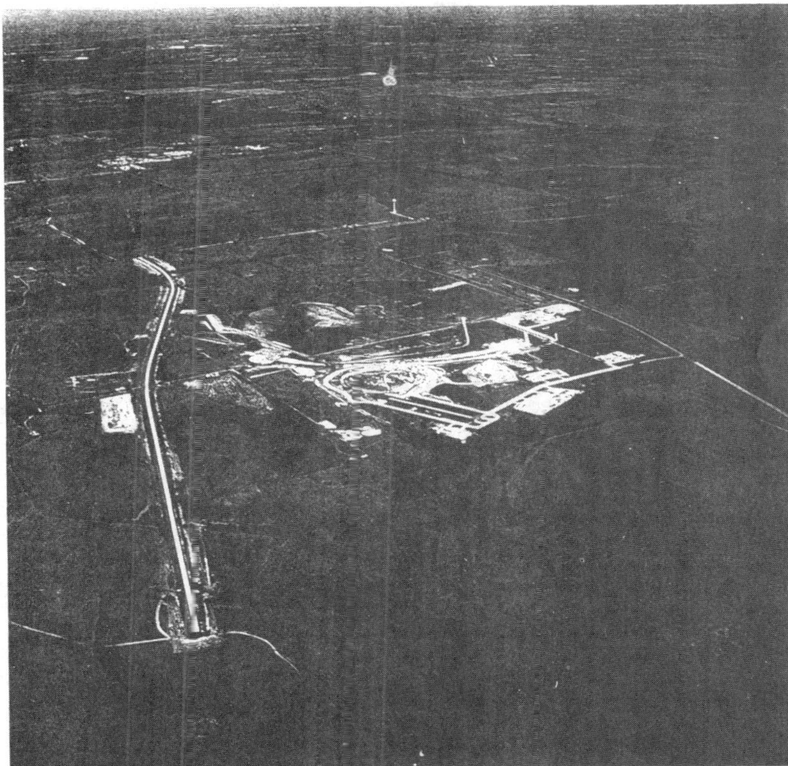

VMS 93-4-121-3
9-27-93
Linac LEB/MEB

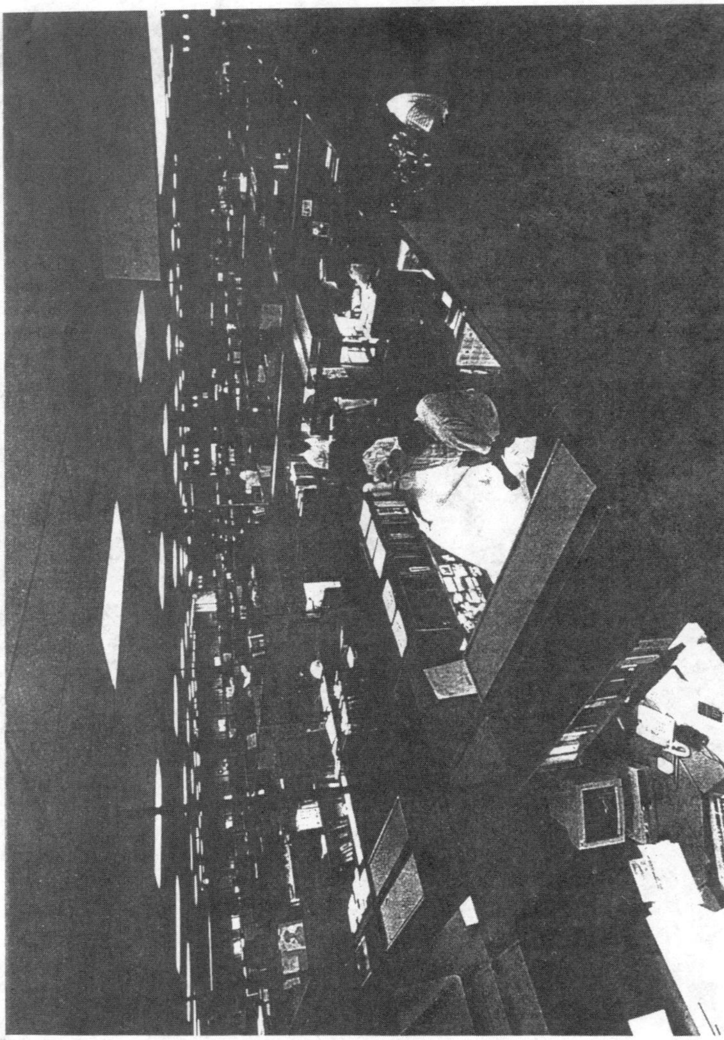

VMS 93-12-13-11
8-18-93
Cubicles/Central Facility

Mr. BOUCHER. All right. Do any of the members who are on the panel with you have any statements to make at this time?

Dr. KREBS. No, sir.

Mr. BOUCHER. Well, thank you very much for that presentation and for your appearance here this morning.

Let me begin by asking you about the plans the Department of Energy has with regard to preserving for later possible use the off-site assets, the assets of the project that were created, accumulated beyond the Waxahachie, Texas site.

It's our understanding that there are advanced detectors, various aspects of the magnet research that has taken place at university campuses and, perhaps, at national laboratories that could potentially be useful in the event that the United States decides to participate with CERN in a high energy physics upgrade for the Large Hadron Collider.

I would like your comments on the extent to which you think those assets might be usable for such a venture, and assuming that you believe they are, what steps the Department of Energy is taking now to preserve those assets for that potential use.

Dr. KREBS. There are two aspects to an answer for that question. In the event that we would go forward in a collaboration with CERN, the capabilities of the people that were developed at the laboratories and at other universities will be—would be able to make a contribution to the effort that we bring to bear in working on the Large Hadron Collider at CERN.

However, in terms of our plans for taking that into account, we are—we have put in place the Drell Subpanel and we are waiting for the results from the Drell Subpanel not only to indicate what the desirable future for the Department's High Energy Physics and the Nation's High Energy Physics Program might be but also to motivate that in the light of the determination not to go forward with the SSC.

The other aspect is that within the context of closure and termination of the SSC we have undertaken to support certain technical completions that involve the capabilities and support the capabilities at both other laboratories and in the universities. And so that is taking place within Fiscal Year 1994—now.

Mr. BOUCHER. All right. Does the plan that you have submitted specify the steps that you would take to preserve the various off-site assets?

Dr. KREBS. The plan that we are required to submit on July 1st would also deal—since it would be submitted after the Drell Subpanel submits its report and after the proposals that are submitted from the expressions of interest should enable us to outline how we would take care of these off-site assets as well as——

Mr. BOUCHER. So we can look forward to receiving your plans with respect to that sometime in the July time frame?

Dr. KREBS. Yes.

Mr. BOUCHER. All right. Let me ask you now about the budget request that you have made for next year. I believe you have requested an additional $180 million for further SSC termination costs; is that correct?

Dr. KREBS. Yes.

Mr. BOUCHER. Let me just review the budgetary history with you, beginning with the Fiscal Year 1994 appropriations conference report which appropriated to DOE $640 million for the SSC termination. It is my information that you also have $95 million in carryover funds from Fiscal Year 1993. That should have made available a total of $735 million for termination of the SSC.

The plan that you submitted to the subcommittee last month indicated a total termination cost estimate of $695 million, which is an amount less than the funds you currently have able, those totaling $735 million.

The revision in that cost estimate which you presented this morning brings the total estimate down by an additional $127 million, and so the revised estimate of total termination costs is $568 million.

So the question is simply this. If you now have $735 million available for termination and your revised estimate shows a total of $568 million in termination costs, why do you need another $180 million?

Dr. KREBS. The request had it—the request for $180 million took into account the fact that we expected to settle in a prompt fashion with the State of Texas, which we believed is in the interest of—the best interest of terminating the project promptly, and also future uses of the assets.

The current revision in the cost, which is not totally reviewed within headquarters at this time does, you know, may require some, you know, relook at that number. But nonetheless, we do believe that we still do not fully understand what the settlement with Texas may be or what the future uses of the assets may entail.

Mr. BOUCHER. So, if I understand your answer, you are proposing these additional funds as a way to preserve assets, primarily?

Dr. KREBS. To maximize the use of the assets.

Mr. BOUCHER. Have you described how those funds would be used in the termination plan which you submitted last month?

Dr. KREBS. No, sir, because we were not—the termination plan that was submitted last month had only to do with the physical activities and the human resources issues associated with the people at the laboratory carrying out the activities associated with the project. That was one part of the termination activities.

The other part has been settlement of the Texas claims and future uses of the assets, and they were not included in that plan.

Mr. BOUCHER. Well, I think we may have something of a problem in terms of our ability now to understand what your plan for the use of this additional $180 million is in view of the fact that we will not know until July what your detailed plans are for preserving the various assets, and yet you have asked in advance of having submitted those detailed plans for $180 million directed toward that very purpose.

It puts us in a somewhat awkward position in terms of our ability to judge whether or not that is an adequate amount of money, whether that would be a wise public investment. Perhaps more funds than that would be necessary. Perhaps less. But given the fact that we will not have that information until July, it is very difficult for us to judge the adequacy of that.

And my question to you is how do you judge the adequacy of it if you haven't devised a plan for preserving the assets as of this date?

Dr. KREBS. I think that part of what we are doing in these next few months is working actually as promptly as we can to, you know, discuss with Texas what, and to come to an agreement with Texas with respect to claims. The whole effort in terms of developing or making the call for expressions of interest should allow us to—and working to assist Texas as it develops proposals that will be fairly reviewed, will enable us to make a better assessment of how much of those funds are actually needed.

Mr. BOUCHER. All right. I have some other questions which I will reserve for a second round and at this time recognize the gentleman from New York, Mr. Boehlert.

Mr. BOEHLERT. Dr. Krebs, it will not surprise you to learn that I am deeply troubled by this. As I understand the situation, we have $735 million on hand to do something that you now conclude will cost us $568 million to do, and yet you are asking for another $180 million on top of that. That is deeply troubling.

How did you arrive at that $180 million figure?

Dr. KREBS. As you may know from your experience and familiarity with the Federal budget process, these figures were established in the fall in probably the November time frame. At that point, in fact, early to, you know—during November, and at that point we did not have the termination plan, even the first draft of the termination plan, available to us. We had not spoken or begun any conversations with the State of Texas, and we had not received even the initial inklings of what sorts of things.

Mr. BOEHLERT. So how did you get the $180 million figure? Sort of a rough guesswork that you just pulled out of thin air?

I mean that's—we're not talking about petty cash.

Dr. KREBS. Some of it was a—had to do with our understanding of what would be available within the overall, the overall targets that were being discussed for the Department of Energy's total budget.

Mr. BOEHLERT. Okay. That's good enough.

Let me ask Mr. Nordhaus a question, if I may. Could you describe DOE's position on the meeting of the 1990 termination agreement with Texas?

And let me say, Mr. Chairman, if I may, I would like to insert in the record at this point an exchange of communications between then Secretary of Energy, Mr. Watkins, and our former colleague Mr. Wolpe, who was chairman of the Subcommittee on Investigations and Oversight in 1991, and I was the ranking member on that. If I may insert these in the record at this point.

Mr. BOUCHER. Without objection, that will appear in the record.

[The documents follow:]

(short)

Congress of the United States
House of Representatives
Washington, DC 20515
July 16, 1991

Secretary James D. Watkins
Department of Energy
Forrestal Building
1000 Independence Avenue, SW
Washington, D.C. 20585

Dear Secretary Watkins:

It has come to our attention that there are conflicting interpretations of the language in the Memorandum of Understanding (MOU) with the State of Texas regarding refunds for contributions to the Superconducting Super Collider (SSC). The MOU appears to endorse merely returning the State's cash contributions that may remain on hand in the SSC Lab's account or any property paid for by the State. However, it has been suggested that some in Texas view this provision as implying a complete refund of any and all contributions made by the State. As you know, the House of Representatives has taken the position that the State of Texas should share some of the risk in building the SSC and voted to bar a complete refund to the State. We would ask that you clarify this issue. How does the Department interpret the MOU on the issue of refunds to the State of Texas?

Secondly, we ask that the Department provide an estimate, broken out by fiscal years, on what proportion of Texas' contribution has taken the form of property that would revert to Texas and how much has been in the form of cash which has supported the various development and managerial activities of the project. In short, how much risk does the State of Texas bear in the SSC project?

Finally, in the event the project were cancelled by the Department of Energy, what is the Department's position on the State of Texas' responsibility for sharing in contract termination penalties? Would the State be expected to share in those expenses, or does the Department believe that the Federal Government alone would bear those costs?

Thank you for your attention to these questions.

Sincerely,

Sherwood Boehlert
Member of Congress

F. James Sensenbrenner
Member of Congress

Howard Wolpe
Member of Congress

The Deputy Secretary of Energy
Washington, DC 20585

August 14, 1991

The Honorable F. James Sensenbrenner
U.S. House of Representatives
Washington, D.C. 20515

Dear Jim:

Thank you for your letter of July 16, 1991, concerning possible refunds to the State of Texas of contributions to the Superconducting Super Collider (SSC) project under the Memorandum of Understanding (MOU) between Texas and the Department of Energy (DOE).

The MOU specifically addresses premature termination of the SSC project in paragraph 10, which provides:

> In the event the SSC is terminated or postponed indefinitely by the DOE or other executive action, or if significant and meaningful appropriations for the construction or operation of the SSC are not provided by Congress in any fiscal year, the DOE will cause the United States Government, to the maximum extent permitted by applicable law, to refund any unexpended cash contributed by the TNRLC [Texas National Research Laboratory Commission] to SSC project accounts and all in-kind contributions of the TNRLC (which for purposes of the MOU is agreed to be any property 50 percent or more of the value of which was contributed by the TNRLC) to the SSC project.

Accordingly, in response to your first question, only unexpended cash and in-kind contributions (i.e., property where TNRLC contributed more than 50 percent of the value) would be returned to Texas.

Your second question concerned the relative proportions of cash and property contributions. The data in the enclosed table show how the total Texas contribution of $875 million toward the Total Project Cost of the SSC is planned to be distributed during fiscal year 1991 through 1996.

This table shows that approximately 52.5 percent of the total Texas contribution would go to fund "other costs" than property for various project development and managerial activities of the project. The remainder goes to fund property and equipment.

In response to the third paragraph of your letter, legitimate termination settlement claims associated with a premature termination of the SSC project could be considered "expended" funds. These funds would be apportioned between TNRLC, DOE, and any other funding participants in the SSC project. Thus, the TNRLC does bear substantial financial risk associated with its participation in the SSC project.

With kindest personal regards, I remain

Sincerely yours,

W. Henson Moore

Enclosure

Mr. BOEHLERT. And, Mr. Nordhaus, before you respond, just let me point out that what the Secretary told us is only unexpended cash and in-kind contributions would be returned to Texas. In other words, the project is terminated. Texas gets land. It gets all the buildings, all the improvements. As we have learned from the Texas newspaper accounts that they have already estimated this project to date has added billions of dollars to the Texas economy. So they have derived a great deal from it.

But it's my understanding, based upon all we have been told from everyone who has testified to this matter that the Memorandum of Understanding between the U.S. Government and the State of Texas, in effect, says we will return the land and buildings, period.

What is your interpretation?

Mr. NORDHAUS. Let me make clear, first of all, that I am a relative newcomer to all of this, have no personal knowledge as to the circumstances that surrounded the negotiation of the Memorandum of Understanding. So I am speaking to it as a relative latecomer, and speaking also in the context of ongoing settlement negotiations and the possibility of litigation. So I want to choose my words fairly carefully.

The State has submitted to us a claim for reimbursement based on its interpretation of the Memorandum of Understanding which, in its view, entitles it to cash reimbursement of all of its contributions to the SSC. This interpretation was submitted to us late last year. We have discussed it with the State. It is not consistent with our interpretation of the MOU.

But I have to emphasize that this is a matter of some fairly sharp disagreement and a question that could potentially end up as a litigation matter between the State and Federal Government and one which in my view is the proper subject for our settlement discussions.

Mr. BOEHLERT. Let me say, Mr. Nordhaus, I am very much on your side. I mean, I think the language is very clear. It seems to me that Texas is trying to make a profit on this project.

Mr. NORDHAUS. It is trying to make a what, sir?

Mr. BOEHLERT. A profit. I mean, I can draw no other conclusion.

Mr. NORDHAUS. My——

Mr. BOEHLERT. The Memorandum of Understanding is very clear and we have been through this time and time again. As the project for some reason is terminated, repeatedly we have heard from the Department of Energy that Texas would lay claim to the land and the buildings, period, and a report.

Let me ask, Dr. Krebs, if you can explain to us why only on-site proposals are being solicited in the call for expressions of interest. I mean wouldn't it—I certainly don't want to exclude proposals from Texas. But why do we exclude proposals from anyplace other than Texas?

Dr. KREBS. The—in some respects, and there may be some comments that Mr. Nordhaus would like to make on this as well, the settlement with Texas and the determination of who owns what—who owns the facilities and how they shall be utilized is tied with, you know, our ability to make a fair determination of what's, you know, what are the useful and appropriate, you know, projects.

Mr. BOEHLERT. But that's obfuscation. I mean, that's a separate issue altogether. I mean, I am asking you a very simple question. Why are we excluding proposals from anyplace else in the world if we truly want to maximize the benefit to the United States of America of our investment to date in this project? Why do we exclude everyplace else in the world?

They can't come from California, can't come from Virginia, can't come from New York, can't come from anyplace in the world—only from Texas. I mean, this sounds like it was developed by the Richards for Governor Campaign Committee.

Mr. BARTON. Would the gentleman yield?

Mr. BOEHLERT. I'd be glad to yield to my colleague. You are not a member of the Richards for Governor Campaign Committee, I know that. [Laughter.]

Mr. BOUCHER. At this point the chair is going to note that the gentleman from New York's time has expired, and with the unanimous consent, without objection, the gentleman will be recognized for 2 additional minutes. But we would like to move along.

Mr. BARTON. Sure. Well, my question to the gentleman would be do you really think you can move the tunnel?

Mr. BOEHLERT. No, obviously you can't move the tunnel.

Mr. BARTON. Okay.

Mr. BOEHLERT. I have got an idea what we can use the tunnel for, but that's not in response to my question.

My question, Dr. Krebs, I think is a legitimate question, and I will abide by the Chairman's wishes and wait for the second round for the question.

But I do think everybody in the whole world is entitled to an answer to a very basic question. Why is the Department of Energy being so exclusive as it deals with this endeavor? Why aren't you saying we will accept proposals from anyone and everyone?

Dr. KREBS. We will accept it from anyone and anyone—anyone and anywhere. However, we are dealing with large-scale facilities that are located, pretty much immobile.

Mr. BOEHLERT. The Chairman says—I'll have another round of questions. The Chairman says my time has expired.

Mr. BOUCHER. The gentleman from Minnesota, Mr. Minge.

Mr. MINGE. I'll pass, Mr. Chairman, to the next.

Mr. BOUCHER. Would the gentlelady from Texas desire to ask questions at this time?

All right. The gentleman from Texas, Mr. Barton.

Mr. BARTON. Oh, jeez! I have never known Members of Congress unwilling to ask questions before. This is quite an event.

Well, let me make a few statements and then I'll ask some questions. When Mr. Cumesty was showing those slides I did—I do want the record to show that those slides were taken when they were building the project. At that point in time you had a workforce of over 2000 people on site. Today, the workforce is approximately 1200; is that the correct? In that neighborhood?

Mr. HALL. It will be 950——

Mr. BARTON. Nine hundred and fifty.

So, if we were to take those same photos today in the central facility that showed the cubicles, those cubicles would be almost totally vacant. If we took the photo in the Magnetic Development

Laboratory that showed the long magnets in kind of the oval shape, there would be nobody. There would be equipment but there would be no people.

So, the Department of Energy has done an excellent job of reducing the workforce on site at the SSC. Hundreds of contracts have been terminated. Numbers of people have been let go, thousands of people. And I think that is important. I would encourage the Subcommittee at some point in time to do a field hearing at the site to see what the status is.

Now, the Chairman asked a very relevant question about how you evaluate the dollars that are going to be needed for termination when you haven't evaluated the alternatives, and that is a valid point. I think these numbers that are being thrown around are not hard numbers, because they don't know yet what they are going to do. And, until they do, it is going to be impossible to give a realistic number to this committee.

I would ask a question of the Secretary. In terms of this latest $127 million reduction, I am told that that is a result of a determination within the Department that no environmental impact statement is going to have to be done. Is that correct or incorrect?

Dr. KREBS. Not explicitly. It is a—the major part of that is associated with a revision of what is required for environmental remediation, and I think Mr. Hall could probably speak to that better.

Mr. HALL. No. The environmental documentation is proceeding now, and there will be a determination as to whether an additional environmental impact statement will be required or what other environmental analysis required.

The biggest change, the largest reduction in that estimate does result from the restoration activities and a big change in how the surface restoration and subsurface restoration would be done.

When the plan was first put together we were trying to get as quickly as possible a baseline for the scope, cost and schedule so we would have something to look at, and then we said we would go back and do an indepth verification of that.

The restoration activities, we utilized the cost estimate that was done by the construction manager that was there at the time, and we have since looked hard, we have got our own contractors in to look at the sites with the representatives from the State of Texas to accompany us——

Mr. BARTON. But my basic assumption is correct, isn't it? You have reduced the termination number because you have reduced the latest estimate on cost of environmental cleanup or restoration?

Mr. HALL. The scope of that, yes.

Mr. BARTON. Okay. That is just the general point that I am attempting to make.

Now, I want to—the chairman also asked about some utilization of assets at CERN, and certainly that is something that needs to be evaluated. Is there any ongoing effort at the Department of Energy to determine what the United States would get in return if in fact we made a decision to use some of these assets at CERN? I mean what do we get at some point in the future if we make a determination to help them and the LHC?

Dr. KREBS. We have—we are not having conversations with CERN at this time. And, of course, if the Drell Panel comes in, and we expect them, in fact, to make some mention of the appropriate involvement with CERN and the Large Hadron Collider, we would then begin, perhaps, to, you know, start that conversation with CERN.

At the moment, as you may know, the Director General of CERN has sent letters, positive letters to both the Secretary of Energy and to the Director of the National Science Foundation saying that he would welcome United States participation, and we have begun a discussion within the Administration amongst the agencies involved as to what that involves.

Mr. BARTON. If it were my——

Dr. KREBS. And, of course, we are interested—we have not made a determination of what we would get——

Mr. BARTON. My strong recommendation to you, Madam Secretary, is if those talks proceed that very soon there be an addendum to them that if we send the kind of money they have asked this subcommittee to send, between $400 million to a billion dollars, that we up front make some counterproposals on what we get in the next stage when the CERN facility has exceeded its useful life.

Now, I want to ask—I wish Congressman Boehlert was still here—Sherry, if you will just wait one minute. I want to take exception to something the distinguished gentleman from New York said about Texas attempts to be trying to make a profit out of this.

I point out that of the Federal dollars expended, at least half of those Federal dollars were spent outside the State of Texas in States like New York where Brookhaven National Laboratory is located, for example.

Point two, the State of Texas money was spent up front at a faster rate than was in the original plan, and the State of Texas taxpayers have really spent at least $440 million. That money is gone. It is not a projection, it has been spent. And all of these termination numbers that are being discussed, there is not one penny so far in any of that to in any way reimburse the State for a good faith effort.

Texas was a partner. The Texas legislature did not pull the plug on the Super Collider. The Texas Governor did not pull the plug on the Super Collider. The U.S. Congress did, and that may have been a wise decision. I think not. You think it was. But the State at some point in time is certainly entitled to an honest debate on that. And to think that we are trying to make a profit is simply counterproductive to an honest dialogue.

Mr. BOEHLERT. Would my colleague yield on that point?

Mr. BARTON. You yielded to me. I will be happy to yield to you.

Mr. BOEHLERT. I would agree with you in terms of the expenditures, but the question is what are you owed in return. We entered into an agreement, the United States Government and the State of Texas, where we both were at risk. That agreement, unfortunately, had to be terminated because the taxpayers' interests were not being served.

Mr. BARTON. Well, that's an editorial comment.

Mr. BOEHLERT. So the question is now what is owed to the State of Texas, and I would suggest that Governor Richards is doing her job, obviously. She's got the hand out and wants as much as she can possibly get.

But, as I pointed out, in rereading the Memorandum of Understanding what we should give to Texas is all the land and all the buildings, and agree with the Dallas newspapers that say it has already added more than a billion dollars to the Texas economy, and say you have got a pretty good deal for your $400 million.

Mr. BARTON. I just want to thank my distinguished colleague. But we need to be up front. We can be rhetorical and we can be editorial, but we need to be specific on the facts of the money spent, whether it was spent and to what purpose, and it is so far in the termination plan not one dime to defray the real dollars that have been spent by the State of Texas.

I will yield back to the chairman on the assumption that we are going to have a second round.

Mr. BOUCHER. We are going to have a second round, and the chair thanks the gentleman for his questions.

The gentlewoman from Texas, Ms. Johnson?

Ms. EDDIE BERNICE JOHNSON of Texas. Thank you, Mr. Chairman. I apologize for having to leave, but you know, with the theory that all politics is local, I had a lot of constituents out there to respond to.

I want to ask several questions. I have been involved with the Super Collider since before it became a part of Texas. Having served in the Texas Senate at the time, I was one of the people that helped to pass the constitutional amendment and worked all over the State to get it—influence the vote of the voters of Texas to support it, and it is and has been a partnership. That was the attitude that we had.

And it was because of that partnership that influenced some of the decision as well as having the appropriate available land, or it being made available for this important project.

I need to ask a number of questions, and hopefully, I will put them out and you can respond to them.

If it is determined that there will be some activity with CERN, knowing full well that I think this type of research has to go on somewhere, how much investment is expected of the U.S. and how many jobs will be in the U.S. because of that? And how can this investment be best made with another country with the taxpayers' dollars?

The other question—that is a general set, and the other set has to do with the investment that has been made at that site. It seems to me that the kind of environmental impact, whether it be positive or negative will depend a great deal on how that site is treated, whether there will be any utilization of that site.

If it is just left like it is at the present time, being such a large tunnel it likely will fill with water. And if you have already addressed this, I apologize, while I was out. And with nothing going into that, it is clear that it could very well go into the water in that particular area and could contaminate the entire area.

So, if there is an area or a move or a plan to address this area, I would like to hear about it, and the cost of it.

The third set of questions, do you feel that it is feasible and what all will it take to manufacture the magnets at that site and sell them to CERN or any other Superconducting Super Collider project throughout the world? I am especially interested if that investment is going to go outside the country.

Dr. KREBS. Let me—I think two of these questions I should address to the extent that I can at this time. The third I think is something that Mr. Hall should address.

With respect to the question of investments at CERN, at this time I don't believe we have a dollar figure established per year or total of what an involvement with CERN would be if we were to participate in the Large Hadron Collider. To some extent I think that dollar figure and that investment is part of making sure that we get a good deal, so to speak, with CERN.

With respect to jobs in the United States, one of the things or benefits to the United States of participation in CERN, I think that comes in two ways. First of all, we already have a few hundred people who are at CERN on a regular basis. Not permanently, for the most part, but they spend a good—a fair number of months a year there. We don't envision seeing a substantial increase if we participate—in that number, if we participate at the Large Hadron Collider.

However, it's obvious to me that our participation at CERN does benefit the young people and our students and, you know, who pursue their degrees there. To some extent I think it can be shown that students who pursue their degrees at CERN then become—you know, come back to the United States generally and become professors of physics or in some respects are involved in the American science and technological effort.

Ms. EDDIE BERNICE JOHNSON of Texas. Let me interrupt there. How many of the students involved now that involves also the Historically Black Colleges that's involved with the Super Collider? Or was involved with the Super Collider?

Dr. KREBS. At CERN?

Ms. EDDIE BERNICE JOHNSON of Texas. Before Mr. Boehlert helped it die.

Dr. KREBS. Ms. Johnson, I am aware that there were a number of HBCUs who were involved in the SSC. I can't give you precise numbers, and if you wish me, I would be happy to provide that for you off line.

[The information requested above follows:]

There were 15 Historically Black Colleges and Universities involved with the Superconducting Super Collider project. To the best of my knowledge, none are presently working at CERN.

But I realize the impact, in fact, that the cancellation of the SSC had on young minority and women students because they had been attracted in some numbers to high energy physics because of the future that the SSC offered. So I am aware but I don't have the precise numbers.

The other element that I would like to address, that I think is appropriate for me to address is something that you talked about, the feasibility of manufacturing the magnets for the Large Hadron Collider at the site and possibly selling them to CERN, I think that that is something that we need—such a project may be an outcome of this process that proceeds under the Agreement in Principle. If that is proposed, we will evaluate it in the context of, you know, the other scientific needs of the high energy physics program and other programs that we may have.

If that—there is no question that the magnet development capability of the U.S. would probably be of interest to the Large Hadron Collider whether it is in Texas or elsewhere, because in fact some of that capability is human capability that's not just at, you know, involving the Texas site. And I think that would also be a topic for discussion with CERN if we were to enter into any kind of conversation with them.

Now, I think Mr. Hall should address the environmental issues associated with the tunnel.

Mr. HALL. As far as the final treatment of the tunnel, we have an environmental analysis underway now looking at that. We have inquired of the Bureau of Reclamation and the Underground Technology Advisory Panel that has advised the project on the tunnel issues in the past.

The aquifer there is located far below the tunnel right now. The water that would get into the tunnel would be surface water that penetrated the rock and some condensation that could get in there. We have been advised to this point that probably bulkheading the tunnel and letting it fill with water over a period of time is an acceptable way to deal with it.

We know of no contamination there that would migrate into groundwater. So we feel that right now, pending further environmental analysis, that is an acceptable way to deal with the tunnel and avoid any contamination of the groundwater.

Ms. EDDIE BERNICE JOHNSON of Texas. Okay. So you are saying that there is no possibility of contamination with groundwater if the water is allowed to accumulate in that tunnel?

Mr. HALL. That is correct.

Mr. BOUCHER. The gentlewoman's time has expired.

The chair now recognizes the gentleman from Texas, Mr. Johnson.

Mr. SAM JOHNSON of Texas. Thank you, Mr. Chairman.

Eddie Bernice, I appreciate you talking about the Texas legislature because I too was there when that occurred. In my view, we had an agreement with the Federal Government that this plan was going to be followed through to the end, and it would have been a marvelous project. I am surprised that we are even thinking about putting it in Switzerland when we ought to have it in the United States.

But notwithstanding, I would like to ask a question. How did you come up with the $127 million reduction? I know Joe Barton asked you that question, but I would like you to be a little more specific on how you developed the number, especially in light of if it has to do with environmental reduction costs when we just heard the statement that you haven't even developed a full environmental study on how to cope with the issue yet.

Mr. HALL. I might—I want to say that there is a mitigation plan. When the original environmental impact statement was done there was a mitigation plan that was a part of that impact statement. We are proceeding to execute that mitigation plan now.

We have changed the scope of the remediation somewhat, but I want to make it clear we are not—we are doing all the environmental work that is required. We are really not eliminating any environmental work. That is all going to be done. All the restoration will be done. The wetlands will be replaced.

Primarily what is changed is a reduction in the cost of that work, a rather dramatic reduction in the cost of the work.

Mr. SAM JOHNSON of Texas. How can you determine that if you haven't figured out what you are going to do yet?

Mr. HALL. Well, we know, for example, the tunnel, which is one of the biggest changes, to be—the estimate that was in the February plan provided for an engineered fill of the tunnel, actually packing the tunnel with dirt. And since that time—that was what the plan was when I got there.

We have looked at that. That practically does not make any sense. There is no requirement to do that. We feel it would even be dangerous to do that. So what we have done is taken another approach, which is fill in the shafts that lead to the tunnel, bulkheading the ends, and then over time it would probably fill with water, which is an acceptable approach based on the advice we have gotten so far.

There is a tremendous difference in how much it costs to do that versus fill the tunnel with dirt. So that's—we don't feel there is an environmental issue there, simply how you go about doing it.

Mr. SAM JOHNSON of Texas. Well, there certainly is. Bernice brought it up. There is waterfill and it's going to contaminate something.

I don't see, though—you still haven't answered my question as to how you can reduce costs simply by changing that when you don't know what the cost is yet. You haven't even estimated the cost on that environmental, have you?

Mr. HALL. Yes. That is the process we have been going through is going around to all the sites that have to be restored, developing the scope for that and estimating that work.

Mr. SAM JOHNSON of Texas. And you have a final estimate right now?

Mr. HALL. We have a——

Mr. SAM JOHNSON of Texas. Yes or no.

Mr. HALL. We have a government estimate for doing that work.

Mr. SAM JOHNSON of Texas. What does that mean?

Mr. HALL. It's not a final estimate.

Mr. SAM JOHNSON of Texas. What does that mean?

Mr. HALL. That means that is the baseline that we think it should take to restore the site. When we get the construction contractors on there to do the actual work, they will do their own estimate of the work. We will then look at that against our estimate to see if we think it's reasonable.

Mr. SAM JOHNSON of Texas. Yes. But you come through with a finite number of what you can reduce costs by and yet you are telling me you don't have that number developed yet.

Mr. HALL. We have—we think we know what it is going to take. We have put an estimate together to do that. We felt it was very important to have a baseline cost for that restoration activity so we can track it as we go along and know when there are changes what those changes are and how it is going to impact the project.

Mr. SAM JOHNSON of Texas. Have you all ever done anything of this magnitude before?

Mr. HALL. This work is—yes, we have. Much greater than this.

Mr. SAM JOHNSON of Texas. How accurate have your estimates been in the past?

Mr. HALL. Well, for uncontaminated work, the estimates are accurate. The problems we have had in the past, particularly in the environmental restoration work we do at Oak Ridge, is that there can be uranium and other toxic materials in the dirt that you don't know until you get in there and start working in it, and that can change, as you know, dramatically.

But we are not working with that here. We don't have that problem. This is rather a straightforward restoration work.

Mr. BARTON. Would the gentleman yield?

Mr. SAM JOHNSON of Texas. Yes.

Mr. BARTON. I would just like to point out that I think it is progress that they have gone from trying to fill the tunnel in to not fill the tunnel in. I think they are to be commended for making that decision.

And I would agree with the gentleman from Texas that they still don't have a handle on the numbers, and until they decide what alternative, if any, they are going to use the tunnel for they will not.

But I think the point you make is well taken. That they really can't give us a hard number. If this gentleman had to swear on a Bible and his oldest child would be shot if he lied, he would tell you the truth, that they don't have a handle on the numbers. Now, I have spoken with him in private on a number of occasions and he is much more, shall we say, expansive when he's not testifying before the subcommittee.

Mr. NORDHAUS. Mr. Barton, could I give a little perspective on this from my vantage point as a lawyer? And that is that I think Mr. Hall is taking a careful look at this from a technical point of view, but one of the things we will have to do before making any decision on environmental remediation is to complete our environmental review under the National Environmental Policy Act, and neither he nor I would want this discussion to be read as prejudging any determination we would make ultimately on what has——

Mr. BARTON. Spoken like a true lawyer.

Mr. NORDHAUS. I did want to state that for the record.

Mr. BARTON. A good lawyer, I should say.

Mr. SAM JOHNSON of Texas. Well but the whole point is, I think, is how can you tell us you don't need $127 million when you don't know. And I think it's wrong for you to come out with a specific amount, reduction like that, without knowing all the answers.

Let me ask you one other question. You stated that there was absolutely no Federal money involved in CERN. Is that true?

Dr. KREBS. No, sir, that is not what I—that is not what I meant to state or meant to give the impression of. In fact, we have an active collaboration with the current activities at CERN. What is being discussed for the future is whether or not we would participate in a new facility that they are proposing called the Large Hadron Collider.

We currently, both the Department of Energy and the National Science Foundation do support the activity, the work of some of our scientists at CERN currently. The issue is what do we—you know, what is our future participation?

Mr. SAM JOHNSON of Texas. Is our participation mostly through universities and grants?

Dr. KREBS. We also—the Department of Energy supports some of its laboratory scientists, its national laboratory scientists at CERN as well, but primarily—it's primarily university participation.

Mr. SAM JOHNSON of Texas. Thank you.

Thank you, Mr. Chairman.

Mr. BOUCHER. Thank you very much, Mr. Johnson.

We will now proceed to a second round of questions and the chair will take a few minutes to propound questions about the proposed new facility in Texas.

Dr. Krebs, we note with interest the Agreement in Principle between the Department of Energy and the State of Texas that suggests a new facility for high energy physics in Texas. It is our understanding that the annual operating budget for that facility might be somewhere between $20 million and $40 million.

And my first question to you is whether that annual operating budget would derive from the Department of Energy's basic program for high energy physics?

Dr. KREBS. I would like to deal with the—I would like not to accept the basic assertions that you have made.

Mr. BOUCHER. Fine. Correct the premise in any way you desire.

Dr. KREBS. I think that what we are involved in now with the State of Texas is a process that first assists them in developing proposals or project definition studies that will be utilized to the maximum, you know, the best, the assets that are at the site. We will then—our understanding is that these projects will be assessed and evaluated as to whether or not they may meet the scientific directions of the Department of Energy or perhaps some other agencies as well.

These proposals are not necessarily restricted to high energy physics. So—or at least we don't expect them to be. And, in fact, you know, at the extent that we have had any indication from communications of various types we would expect them to range quite broadly. And so—but we see this as a commitment to a process that enables project definition and then a fair evaluation of the contributions.

Mr. BOUCHER. So you are not as far along as suggesting that a new facility for high energy physics be constructed at that site.

Where did the figure of $20 million to $40 million per year in annual operating costs come from? Did the Department supply that figure?

Dr. KREBS. I believe that that figure is coming from some reports and communications from the State of Texas.

Mr. BOUCHER. All right. But that is not a figure that the Department has calculated nor is prepared to estimate as the annual cost for such a facility?

Dr. KREBS. No, sir.

Mr. BOUCHER. All right. Now, you provided some very informative remarks with regard to the process.

Mr. BOEHLERT. Mr. Chairman?

Mr. BOUCHER. Well, I'll call on the gentleman momentarily. Let me ask this question first and then I'll be happy to recognize the gentleman on his own time.

You have provided some very informative comments with respect to your Memorandum of Understanding with the State of Texas concerning a possible new project with that State. My question to you is this.

Would you inform us of the means by which the Department of Energy intends to assure that any project with the State of Texas that will enjoy Federal funding will receive adequate scientific comment and adequate scientific review, and that it will be weighed in competition with other projects that also would compete for Federal funding in whatever field is chosen?

What assurance can you provide us that that review and that scientific comment will be provided prior to any undertaking by the Department with respect to a project in Texas?

Dr. KREBS. Mr. Nordhaus may want to comment on this, but as I understand the Agreement in Principle, we have to—the State of Texas and the Department of Energy have to jointly agree that this is of mutual benefit, and we would need to bring those—you know, our method of jointly agreeing certainly from the Department of Energy would involve an assessment of the contributions to the scientific directions of the Department.

Mr. BOUCHER. So what you are saying is you would go through your normal process of scientific comment and review in order to determine scientific merit before any decision is made on the part of your Department to recommend funding for and support of that project?

Dr. KREBS. I think that that would be involved, yes, before we would bring it forward.

Mr. BOUCHER. Well, a simple yes would serve the purpose here.

I think one thing that we would insist on from our perspective is that the Federal Government's normal process of scientific review be undertaken before any new project is launched, and I would very much hope it would be the Department's intention to observe normal processes with respect to any new scientific project we undertake.

Dr. KREBS. A part of what's, you know, involved here is when you—in terms of my hesitation to say a simple yes, is normal process, you know. There's a difference between normal process for grants or projects that are at the $100,000 and $500,000 a year level and the process that might be involved for a substantial commitment, because then what weighs is not simply a normal process of, shall we say, peer review but there are, you know, there are various kinds of balances that get made within the Department relative to—you know, from one program to another, not just within the scientific programs, but perhaps the applied science and technology programs, and that's what, you know, if we're talking about a substantial investment, those kinds of balancing judgments are going to be made as well.

Mr. BOUCHER. Well, I thank you for the answer. I would assert again that we will be watching very closely the process from our end to make sure that an adequate level of scientific review does take place before any Federal investment is made.

Would you care to—all right.

Dr. KREBS. All right.

Mr. BOUCHER. One other question I would ask is this, or perhaps statement to make, and if you would like to respond, you may. We'll be very interested to receive in the May time frame the Department's plans with respect to the preservation of assets emanating from the terminated SSC project which in turn would reflect on the adequacy or lack of adequacy of the $180 million special funding request that DOE is making for fiscal year 95 for that purpose.

We can't make any evaluation at the moment with regard to whether those funds are appropriate or not because the plans are simply not before us. So, we will await with interest your submission of those plans during July.

The gentleman from New York, Mr. Boehlert.

Mr. BOEHLERT. Thank you, Mr. Chairman.

I would suggest that Dr. Krebs is probably an excellent bridge player because that was a magnificent finesse.

I couldn't agree more with the chairman's assertion that we are going to insist on peer review, merit review, merit-based decisions, and I was somewhat bemused by the response to the question.

It seems to me that the Department is hellbent on working out an arrangement with Texas, and forget everything else, we are going to do it. Not if I have anything to say about it, not unless it passes the stringent test that the chairman has insisted it pass.

And I might point out, and I wanted to interrupt briefly the chairman, the $20 to $40 million figure referred to is not just from a Texas report. Secretary O'Leary used that figure herself in testifying before the House Committee on Appropriations.

But let me get back to the on-site proposals and why only on-site proposals are being solicited in the call for expressions of interest. Let me point out in *Inside Energy* recently there is an article that said researchers—headlined "Researchers Say Millions of Dollars in SSC Equipment May Be Wasted," and it quotes a researcher from the High Energy Physics Advisory Panel, Jonathan Dorfman, who says that DOE may be spending millions of dollars on current accelerator construction programs to buy new scientific equipment identical to that now lying idle at the former SSC.

And he went on to say that we're being held hostage by—they are holding us hostage. How do you explain to that—respond to that?

We can get all of the equipment that the people want from Stanford, from other places, and you are saying no, only on-site proposals will be considered.

Mr. NORDHAUS. A couple of—Assistant Secretary Krebs asked me to respond because it implicates some of our concerns with the negotiation with the State. One of the—there are a couple of things here.

First of all, with respect to the proposals themselves, we have—the reference to the site is proposals to utilize assets at the site. Proposals can come from anywhere. They don't have to come from Texas. The ultimate utilization does not have to be at the site.

Secondly, with respect to whether we can in fact make this equipment, to the extent it is moveable, available elsewhere, one of the key issues that we have to work out with the State is the question of who owns it. And I think as you indicated in your earlier discussion, the reading of the Memorandum of Understanding that you had, and which is widely shared within the Department, would end up with the State having an ownership claim to much of this equipment.

Mr. BOEHLERT. Oh, I understand that. But my whole point was there is a great deal of interest beyond the borders, and that is critically important.

Dr. Peoples, you are getting by too easily. I would ask you to describe the current State of high energy physics funding. I think that's a very important subject, the current state of high energy physics funding.

Could you flip the mike on, please?

Dr. PEOPLES. There is a strong belief within the community that the level of funding for the existing high energy physics program is about 20 percent less than it was or less than is desirable.

Mr. BOEHLERT. So the current state of high energy physics is not very favorable.

Dr. PEOPLES. It is not favorable.

Mr. BOEHLERT. Well, we have other accelerators. Can you give me an indication of how they will be utilized next year given the current sorry state of high energy physics funding?

Dr. PEOPLES. The expectation is that at Stanford Linear Collider and the Fermilab Tevatron there will be a reduction in utilization of 30 to 40 percent, and at Brookhaven's Alternating Gradient Synchotron it is more severe because their budget is smaller.

Mr. BOEHLERT. So we are really suffering and we could certainly use some additional money to have those accelerators up and running on a more—well, with a better schedule? Let's put it that way. A hundred and eighty million dollars might help out, wouldn't you suggest?

Dr. PEOPLES. Well, clearly you have offered a nice bauble. But I think less—in this time when people have to be economical, less would be better.

Mr. BOEHLERT. All right. Time's up says the chairman.

Mr. BOUCHER. The chair thanks the gentleman, and recognizes the gentlewoman from Texas, Ms. Johnson.

Ms. EDDIE BERNICE JOHNSON of Texas. Thank you, Mr. Chairman.

I need to just ask one more round of questions involving the environmental impact. Being told that the water accumulation would not contaminate anything to do with the groundwater, I am really wondering whether or not having the tunnel filled with water will do something to the surface, would cause the ground to collapse or—I just can't imagine a tunnel that large—I have been in that tunnel—accumulating water and holding it and nothing changing in the land around, under it or on top of it or somewhere.

What do you—from a scientific point of view and an environmental point of view, what do you feel would be any changes, if there will be any, other than the state that it's in currently?

Mr. HALL. As you know, much of that tunnel is lined and that is not going to change or collapse at all. The unlined portion, the advice we have had is by permitting it to fill with water, equalizing the pressure will—is an acceptable way to preserve the tunnel, to keep it from collapsing, so that if you ever wanted to some time in the future you could pump the water out and utilize it.

One could put sump pumps in and keep the water out at all times, and we are—as a part of the environmental documentation that Mr. Nordhaus referred to, before we make any final decisions on how the remediation will take place, those alternatives and options will be analyzed and looked at, and we will have additional experts help us on that.

But certainly we are not going to contaminate the groundwater. I mean that is not going to happen. Whatever it takes to make sure that doesn't happen I can assure you is our goal.

What we have been advised so far, the options we have looked at and we have priced out for the purpose of the termination plan, it does not appear that we will have any groundwater problems. But we will continue looking at that and the final options will be analyzed.

Ms. EDDIE BERNICE JOHNSON of Texas. Are you getting public comment? Or who's advising you now?

Mr. HALL. When we finalize the restoration plan, get the final restoration plan, we will have public comment. We'll hold a public meeting, make it available publicly and have a public discussion of it.

Ms. EDDIE BERNICE JOHNSON of Texas. Um-hum.

Mr. HALL. So that will happen, yes.

Ms. EDDIE BERNICE JOHNSON of Texas. Thank you. Now, the other question I had, when the chairman spoke of the new facility, are we speaking of a new facility or the use of an existing facility?

Dr. KREBS. We would be talking about maximizing the use of existing facilities. Indeed, we are talking about maximizing facilities.

Ms. EDDIE BERNICE JOHNSON of Texas. Thank you.

Thank you, Mr. Chairman.

Mr. BOUCHER. The chair recognizes Mr. Barton.

Mr. BARTON. I thank the chairman.

I have some specific questions that I postponed in the first round that I want to ask on some of the alternatives.

The Secretary of Energy has received a letter from the Governor of Texas in which the Governor of Texas, based on a recommendation of her Advisory Committee which received a recommendation from the National Academy of Sciences panel, headed by, I think, a Mr. Robert—a Dr. Smith, that $44 million be set aside immediately to maintain a core group of scientists, I think the number is 125, for them to take the next year or year and a half to actually study some of these alternatives.

It is my understanding that this recommendation is within the spirit and the letter of the law on maximizing the assets and that the Secretary has the discretion to accept that recommendation immediately.

Are you familiar with that recommendation? And if you are, could you advise the committee as to what your recommendation would be to the Secretary as to whether to accept or reject it?

Dr. KREBS. I am familiar with the letter that came to the Secretary. My understanding is that the Agreement in Principle provides for a process within which the Department of Energy can work with the State of Texas to examine those, or to develop those—to assist the State in the development of those proposals.

Mr. BARTON. Okay. But, Madam Secretary, this is a specific recommendation. The National Academy of Sciences has recommended that of available funds $44 million be set aside immediately to maintain a core group of scientists at the site to review these alternatives. That is not a recommendation for an alternative. It is a recommendation for the expenditure of termination funds in which to develop alternatives in the next year and a half. If you don't maintain the people to do the evaluation, an honest and fair evaluation cannot be done.

Now, this letter was sent to the Secretary of Energy over a month ago, and to my knowledge—and I may have imperfect knowledge, I will admit that—but to my knowledge, no official response has been obtained by the Governor on this recommendation.

And Dr. Peoples I know is aware of the recommendation. He may—I don't know if he is in the chain of command to evaluate it, but he could certainly comment on it.

Mr. NORDHAUS. Could I make two preliminary comments? Then I will turn it over to Assistant Secretary Krebs. First, the appropriation is available for purposes of hiring consultants and others to assist——

Mr. BARTON. Mr. Counsel, we are not asking to hire consultants. This recommendation is not hiring anybody.

Mr. NORDHAUS. Yes.

Mr. BARTON. It's expending funds available to maintain people that are already there.

Mr. NORDHAUS. Let me put it another way. It can be used to continue to employ contractor personnel for purposes of formulating the report that is due in July. That is the first thing.

The second thing is that the Agreement in Principle does provide for some level of support to the State to assist the State in developing its own proposals. The level of support is something that is now being determined, and I think Dr. Krebs can address that.

Dr. KREBS. First of all, a matter of clarification. We are talking about two different communications. I mean the National Academy made some recommendations to the Hay Commission. The Hay Commission forwarded their interpretation and, if you will, a gloss on those recommendations. It's my understanding that the National Academy made a recommendation that $20 to $30 million might be appropriate to sustain capability on the site.

Upon receiving it then, the Hay Commission forwarded, you know, some additional consideration on the recommendations of the National Academy. Upon receiving——

Mr. BARTON. I agree with all that.

Dr. KREBS. Okay. I just want to make—but I want to make, you know, a clarification.

Mr. BARTON. I want an answer. Okay?

Dr. KREBS. It is with—right now we are, within the framework provided by the Agreement in Principle, we are working with the State of Texas to determine what—how they want to move forward with developing proposals, what it will take, how many people may be required to be retained at the laboratory.

Mr. BARTON. Madam Secretary, here is my point. Both the proponents and the opponents of the Super Collider agree on one thing, and that's that it is time to really try to look at alternatives so that we can get these numbers more definite. Now, Sherry Boehlert and I don't agree on what has happened, but we both agree that we need to have some meat to evaluate.

Now, your Department has got a specific recommendation on the table from the State of Texas to put aside some of the money that was appropriated to maintain a core group of people that can evaluate these alternatives. Now that didn't take a rocket scientist to say, "Did they ask for too much, too many people, too few people." But, you know, make a decision and then start looking at alternatives.

To my knowledge, no one in the Department of Energy has been tasked seriously to look at alternatives. Now, the Governor of Texas empaneled an Advisory Committee of very distinguished people, Medal of Science winners; past presidents, University of Texas, University of Rice; physicists—good people, and they have been meeting for 2 months, and they have received over 70 written alternative proposals, and they have narrowed that down to three or four, and your Department has agreed in some agreement that was made a couple of weeks ago to take some of those recommendations and seriously evaluate them.

Now, you told the subcommittee chairman that you are going to do that, I think, by May. Now, I may have missed that. It is time to do that. It is time to do that.

And we can talk all day about how you are going to evaluate the environmental impact of water seepage and stuff, but, you know, I am not trying to run your Department, and I am not even trying to—I wish I had the power to rebuild the Super Collider project, but I admit I don't.

But at this point—I mean at some point in time I may have to go to battle with Congressman Boehlert. But I want somebody in the Department of Energy, and you are the distinguished representative of the Department, to get off your tail and look at some of these alternatives and either shoot them down or build them up. And do the peer review that we have asked to be done, and bring it forward with the numbers and let this subcommittee and full committee do our job.

Mr. MINGE [presiding]. The gentleman's time has expired.

Mr. BARTON. Well, I would like for her to give a response. And I don't know what constraints she's operating under today but she hasn't given a direct answer to this direct recommendation that the Governor of Texas made to the Secretary of Energy.

Mr. MINGE. Can you answer this quickly, and then we will go on to the next witness, the next speaker.

Dr. KREBS. I am sorry. It took us several months to get the request with or the expressions of interest going and out on the table, which we put out there in conjunction and with the agreement of the State of Texas. We needed—we wanted to put that expression of interest request on the street with the agreement of the State of Texas. Within the framework of the Agreement in Principle, we have got that out there.

Although the letter to the Secretary in a sense can be considered a proposal, in order to enable the Department of Energy to respond to it as a proposal we needed to work with the State of Texas to get that into a frame that, in fact, overall and in terms of our procurement process, we can respond in a productive and effective way. And we are now within the Agreement in Principle, moving forward on that.

We have had conversations with the State of Texas to, you know, to put ourselves in a position where we can receive and respond promptly and effectively to what we would call project definition studies from the State of Texas.

Mr. MINGE. The chair recognizes Mr. Johnson from Texas.

Mr. SAM JOHNSON of Texas. Thank you. I appreciate that.

I would like to carry on with what Mr. Barton started with a little bit. Do you have people in the Department now, and if so, how many, looking at proposals and/or have you received any at all?

Dr. KREBS. We just went out a few weeks ago with what we call a request for expressions of interest. Those expressions of interest we expect to receive by April 15th So we have not at this time received any explicit proposals that we would recognize within our procurement process.

Mr. SAM JOHNSON of Texas. So you don't recognize the Texas proposal as a proposal, is that true?

Dr. KREBS. We are helping them to develop that proposal, to develop their concept into a proposal that we can procure to.

Mr. SAM JOHNSON of Texas. Do you have a specific division in your Department assigned to take care of this project when it starts developing, or are you going to assign people to that when you get proposals?

Dr. KREBS. We have a group within my office that will manage the review and—the review process. To some extent we—and in conjunction with the State of Texas, because under the Agreement in Principle the review and evaluation process of these proposals must be mutually developed.

We expect proposals of varying natures that are not necessarily ones that simply respond to scientific opportunities but also might be economic development opportunities as well. And we will need to work in a sort of cross-cutting way within the Department to assist in the review of them.

Mr. SAM JOHNSON of Texas. Well, do you anticipate a timetable for completing that effort? I mean if you are saying you are going to have April the 15th some proposals to consider, why is it going to take you till July to get an estimate to this committee on what the costs are going to be and what is going to happen?

Dr. KREBS. We are working with the State of Texas. There are two activities going on right now. We are working with the State of Texas to find a way to turn the letter from the Governor into a proposal that we can support, and we have been working with representatives from the State of Texas and we would expect to have a proposal from them within the next few days, and we think that we can, you know, provide funding for that activity at the site within another week or so beyond that.

But with respect to expressions of interest, where we expect many more proposals, we are talking about getting money out probably in the May time frame. In mid-May? Mid-May, to the expressions of interest.

Mr. BOEHLERT. Would my colleague yield for a moment?

Mr. SAM JOHNSON of Texas. Sure.

Mr. BOEHLERT. Is this going to be peer reviewed, the various proposals? The various proposals, are they going to be peer reviewed?

Dr. KREBS. These are proposals for project definition studies, and then when these project definition—when the project definition studies come in that is the—they will then be full, you know, technically and scientifically complete enough so that we can subject them to a peer review or to the peer review process within the Department.

Mr. BOEHLERT. Well, that's something I think that all of us up here agree on. The search and the quest for peer review.

Ms. KREBS. Um-hum.

Mr. BOEHLERT. And I wish I had a lot more confidence than I do in what's happening down there.

But I will thank my colleague for yielding.

Mr. SAM JOHNSON of Texas. Thank you, Sherry.

Let me ask you a different question, if I could. You know, people say that high energy physics funding is 20 percent less than optimum. Can I ask you what you intend to do with the SSC funding that was out there that is being reduced, and is it going into small science or is it going into high energy science? I mean are we reducing our total overall science budget?

Dr. KREBS. Within the Department of Energy, I mean I need to give you perspective here. It is not going into small science. The Department of Energy's budget went from $18.5 billion in 1994 to $18 billion in 1995. The overall budgets for my programs that I am responsible for went from $3.3 billion in 1994 to $2.9 billion in 1995. So, you know, the level of effort in science within the Department of Energy has not, you know, has not benefited, if you will, from the termination of the SSC.

Mr. SAM JOHNSON of Texas. True. But your reduction in your particular budget is not a 20 percent reduction overall.

Dr. KREBS. No.

Mr. SAM JOHNSON of Texas. Can you say that high energy——

Mr. MINGE. The gentleman's time has expired. There will be another round. If you would like to pursue it then, we could do so, or if you would like to take it off your time remaining on the next round.

Mr. SAM JOHNSON of Texas. Yes, let me do that. If I could just finish the question.

Mr. MINGE. Okay. Sure.

Mr. SAM JOHNSON of Texas. I would just like to ask if the overall high energy physics budget has dropped 20 percent in your Department?

Dr. KREBS. The high energy physics budget absent the SSC is essentially at constant dollars between 1994 and 1995. If you want to talk about level of effort over the last few years, I think when you hear the 20 percent—the way people calculate the 20 percent reduction is that a considerable amount of money for the base, for scientists in the high energy physics community had been carried by activities in the SSC. On top of that there had been some operating activities at the ongoing facilities had also been cut back as a result of expecting the SSC to come on line.

Mr. SAM JOHNSON of Texas. Thank you. I appreciate it.

Thank you, Mr. Chairman.

Mr. MINGE. The chair recognizes the gentleman from New York, Mr. Boehlert.

Mr. BOEHLERT. Thank you, Mr. Chairman.

Dr. Krebs, back in October Secretary O'Leary sent a letter to the chairman and to me saying, in effect, that DOE could not afford another lab. Now, you are entertaining a whole lot of proposals. Given the state of your budget, how could you fund another lab?

Dr. KREBS. I think that that is—there is a challenge if, in fact—but first of all, before I say that, what we are engaged in is a process to determine whether or not there is a way to maximally use the assets of the SSC. Some of that may be within our programs of the Department of Energy, not necessarily high energy physics.

Obviously, those facilities and those activities would have to measure up and be weighed against the, you know, the proposals that are already being considered within our other, you know, the other programs.

And I'd just like to—this is kind of anticipating some of the—I'm going to appear I guess before another subcommittee of this committee on Thursday.

Mr. BOEHLERT. This is just a warm-up for you?

Dr. KREBS. And essentially, you know, despite the fact that I am dealing in the Department within the basic research programs, they are dealing with a constant dollar budget. We do have to think about the future, and we do have to think about new facilities within high energy physics. We are thinking about the next 25 to 30 years. So we are beginning things.

Mr. BOEHLERT. Pardon me, if I may. I mean in view of what Dr. Peoples just told the Subcommittee we don't have enough money to run Fermilab or Stanford or Brookhaven at the optimum level and yet—we don't have enough money to run them, you are just telling us, you are acknowledging that devoid of SSC the high energy physics project is flat. How could we under those circumstances possibly go back on what Secretary O'Leary told Chairman Boucher and me in October, that we just couldn't fund a new lab?

Well——

Dr. KREBS. I don't disagree with you, but what I am saying is that we are not necessarily looking simply at activities that would be within high energy physics.

Mr. BOEHLERT. Dr. Peoples, hasn't HEPAP expressed some real reservations about this?

Dr. PEOPLES. HEPAP is a panel that actually advises Dr. Krebs. But there was a statement by the subpanel chairman, sort of a preliminary report about what future options would be. But I am not aware that Professor Drell said anything about the current base program.

Mr. BOEHLERT. Well, by studying a new lab, he certainly did.

Dr. PEOPLES. Well, unfortunately, sir, I was not present on the Monday of that meeting. I went to the Tuesday of that meeting, because I had to give a talk.

Mr. BOEHLERT. Okay.

Dr. PEOPLES. And I apologize.

Mr. BOEHLERT. Well, we'll wait till we hear from the Monday group.

Dr. PEOPLES. Okay.

Mr. BOEHLERT. All right. Let me ask you a question. This will be my last question, and it's not a big item but it's an item of great concern, and it's directed to you, Dr. Peoples.

We have been led to believe that there are problems at the SSC site with the theft of computer equipment. Could you address that?

Dr. PEOPLES. There has been some disappearance of work—I shouldn't say work stations, sort of MacIntosh-type of equipment, in particular from one building. The total value was about $40,000. It's a small fraction of the total.

Mr. BOEHLERT. I understand that.

Dr. PEOPLES. We have reported it to the proper authorities and it's being investigated. We have increased the security at the laboratory steadily since October.

Mr. BOEHLERT. So there have been no further incidents that you——

Dr. PEOPLES. Actually, most of the equipment that cannot be accounted for is in some sense lost. It's a little hard to imagine how something very big gets lost, but it eventually turns up. One of the pieces that was reported——

Mr. BOEHLERT. Well, there's a big difference between lost and theft. I think you would agree to that.

Dr. PEOPLES. That is correct.

Mr. BOEHLERT. And I can appreciate the difficulty in shutting down a project of this magnitude. But we also have an obligation to protect the assets owned by, depending upon who you believe here, the people of Texas or the United States Government.

But you say increased security measures have been instituted?

Dr. PEOPLES. Yes. The total amount that we are spending on security is somewhat over $250,000 a month. Now, that covers more than the buildings. It covers the construction sites. It covers a large number of things. And we are taking that matter very seriously.

Mr. BOEHLERT. I just want to conclude this on a positive note. Earlier my colleague from Texas, Mr. Barton, made a statement that DOE doesn't have a handle on the numbers. I agree with him 100 percent.

Thank you.

Mr. MINGE. The chair recognizes Mr. Barton.

Mr. BARTON. I thank the chairman and I thank his attendance. I know this is a subject that's not quite as near and dear to his heart as it is to some of the others, and I appreciate his patience.

I want to ask Dr. Peoples a question. He wasn't introduced, but he is the permanent Director of the Fermilab, and he is, I guess, the Acting Director of the SSC Laboratory, so he's double-hatted. Our next panel is going to talk about some of the alternatives.

In your position have you—are you generally familiar with the alternatives that have been reviewed for the SSC, and would you care to comment on any of the specific alternatives that the next gentleman is going to ask be seriously reviewed?

Dr. PEOPLES. I have read the letter of Mr. Jess Hay to Governor Richards and I know reasonably what's in there. That there are sort of three, I'll call them areas of proposals, they are not specific proposals.

Mr. BARTON. Okay. Would you—one is on the utilization of the magnet development magnet—Magnet Development Laboratory and the Magnet Test Laboratory to try to see whether it can be used in a broad way for cryogenics and superconductivity. It wasn't specifically related to high energy physics.

A second one was to use the computing equipment that is at the SSC laboratory and is currently in use as the nucleus of a regional computing center.

And the third one was to utilize the partially completed linac for a medical facility.

So, there is really only one of those three that would touch on high energy physics, and that would be the superconductivity facility. Is that correct?

Dr. PEOPLES. That's correct.

Mr. BARTON. Okay. Now, how would—there were other alternatives that the panel, the State of Texas Advisory Committee has not forwarded to the Governor who in turn has forwarded to this group at DOE that's going to look at them. Are there some high energy physics alternatives that, with your knowledge as a physicist, you would recommend be reviewed in addition to the others?

Dr. PEOPLES. These things called proposals in part were developed by laboratory employees and there are things in there that are high energy physics. I think they would have to be scrutinized.

Mr. BARTON. Well, let me ask the question in a different way.

Dr. PEOPLES. Yes.

Mr. BARTON. Do you think that as part of the review process there should be a review of some use of the facilities for a high energy physics purpose? I think you said yes, but——

Dr. PEOPLES. My answer to that would be yes.

Mr. BARTON. Okay.

Dr. PEOPLES. I mean here are some assets. I think if you want to put them to use you would try to find the most appropriate and cost effective use for them.

Mr. BARTON. Okay. Now, I have two comments and then I'll yield back to the chairman.

We need to be very careful when we—you know, when Congressman Boehlert has talked about the high energy physics budget being underfunded, I would agree with that. The distinction I would make, there is a difference between spending money to warehouse people so that they may have something useful to do in the future and spending money to let scientists do leading edge high energy physics. And one of the reasons that the Super Collider was supported by the scientific community is because there was no question that it would do leading edge high energy physics research.

The existing laboratories at Fermilab, Brookhaven, SLAC, and even CERN, there are still some useful things to be done in the B–Factory is an example of that. But at some point in time, if we do not make a decision, and I know this point is in the future, you can spend all the money in the world, and if they just pay people to sit around, play games on the computer screen, and they are not really engaged in leading edge research, it is not money well spent. That's the first point.

The second point. There has been some attempt, and not necessarily by this panel, to make the case that the assets at the SSC are wasting away. Certainly the people assets are wasting away. But I have been on site numerous times since the project was killed and I can stipulate that the security is much tighter after termination than before in terms of bodies and credentials and checkpoints and doors locked. Much tighter.

And the equipment itself is being well maintained. I mean well maintained. And many of the computer terminals have actually been shrink-wrapped in the warehouses.

I mean you can argue that the equipment should be sent somewhere and used. But I don't think you can argue that the equipment that is there is not being maintained in a fashion that it could be used once a decision is made on the alternative.

Okay. I would yield back.

Mr. MINGE. Thank you, Mr. Barton.

Well, that completes the questions, and we would like to thank the panel for being here, for answering all of these questions and listening to the statements with such patience. And we welcome portions of the panel back later in the week to appear before another subcommittee.

So the panel is excused.

Our next panel consists of one individual, Mr. Shelton Smith, who is the Chairman of the Texas National Research Laboratory Commission of Dallas, Texas.

And I understand, Mr. Barton, that you are familiar with Mr. Smith, and I would defer to you to introduce Mr. Smith to the subcommittee.

Mr. BARTON. Very briefly, Shelton Smith is a distinguished attorney in private practice in Houston, Texas. He agreed to take on the unenviable task of chairing the Texas National Research Laboratory Commission in this difficult period. He has devoted many hours, I can attest to that, on an unpaid basis doing so. He is doing an excellent job. He is operating in a good faith fashion.

And I would tell you and the other distinguished Democrats on the panel that he is a Democrat in good standing with the Democratic Party in Texas and is proud of it, and represents his party when necessary in a very ethical fashion.

And with that I would yield back to the chairman.

Mr. MINGE. Thank you.

At this point, in the interest of time and the hour, what I wish to do is to defer to the other members of the panel to ask any questions. But in the beginning, Mr. Smith, I would like to call on you for your opening statement, and request, if possible, if you can limit yourself to 5 minutes, and the written statement that you have submitted will certainly be accepted and will be a part of the permanent record of these hearings.

So, with that opening comment, I would like to ask you to proceed.

STATEMENT OF SHELTON SMITH, CHAIRMAN, TEXAS NATIONAL RESEARCH LABORATORY COMMISSION, DALLAS, TX

Mr. SHELTON SMITH. Thank you, Mr. Chairman. I appreciate the opportunity you have afforded me to be here today.

As chairman of the Texas National Research Laboratory Commission, I'm really here today in two roles. First, obviously, I am representing the State of Texas. As this subcommittee knows, approximately one of every $4 that was spent on the SSC was Texas money. Thus, as a Texas official I have the legal obligation to my Governor and to my State to ensure that the termination process is accomplished in a manner that is consistent with the congressional mandate to minimize the loss to both the United States Government and the State of Texas.

But, members and Mr. Chairman, I am also here in a second capacity. I am here as your partner. The State of Texas and the Federal Government undertook this project as partners, and I think this subcommittee knows that Texas has throughout the history of this project been a good and a faithful partner. We have always, without fail, done everything we were asked to do, everything we were required to do, and then we have done more.

As your partner in this project I then have the responsibility to work with you and with the Department of Energy to ensure that this termination process is accomplished in a manner that is consistent with the congressional mandate to maximize the value of our joint investment.

Members, our $2 billion joint investment has produced some incredibly valuable assets that clearly have the potential to be of great and beneficial use not only to our Nation and to my State but to the entire human race. I believe that we must work together to maximize the values of these assets, and I want this subcommittee to know that on behalf of the State of Texas and on behalf of Governor Ann Richards we are ready, willing and able to work with you to do that.

The day Congress voted to terminate the project Governor Richards met with Secretary O'Leary and other Administration officials to begin planning the process to orderly terminate the Collider. Since that day almost 25 months ago she has worked long and hard with Secretary O'Leary and other Administration officials in a spirit of cooperation and in a spirit of good faith in an attempt to resolve the many complex issues that are facing us in termination of a project of this magnitude.

As the subcommittee has heard this morning, Governor Richards several months ago created a special advisory committee separate and independent from my State Commission whose sole purpose was to solicit, receive, evaluate and analyze recommendations for uses of the SSC assets. That special committee, chaired by Mr. Jess Hay of Dallas, is composed of some of the most eminent scientists, educators and business leaders in the entire State of Texas. These people come from Houston, they come from Dallas, some of them are Republicans, some of them are Democrats, but they are committed to doing a good job. And I think if the subcommittee has seen their work product, you will agree that in a short period of time they have achieved a great deal.

This Advisory Committee—and I am not going to go through this, Mr. Chairman; it's in my written statement. I think it might be more productive if we just did it on question and answer. A lot of this has been talked about this morning, but the Hay Committee has made three recommendations of which the subcommittee is well aware.

My Commission has formally endorsed and adopted the Hay Committee recommendations and I have advised Secretary O'Leary on behalf of my Commission that we believe that they ought to be looked at and that eventually they ought to be implemented.

There is one thing that is not in my prepared remarks that I want to address. It's an issue that came up this morning. And, if I may, Mr. Chairman, let me just address this, then we will go to questions and answers.

The issue was raised this morning as to whether or not the State of Texas is trying to make a profit off the termination of the Super Collider. Mr. Boehlert is concerned about this, and that is a legitimate area for concern. No State, no individual, no contractor, no company should try to profit off the demise of this magnificent project. And I can assure you, Congressman, that the State of Texas is not attempting to do that.

If we look objectively at what the State of Texas has done and what we expected to get in return, it's very simple. We had agreed, the people of Texas, to put up all the land necessary for this 200-square-mile project and up to $1 billion of our tax dollars to help finance this project.

There's no secret that we expected some things in return, and there is no secret that those things have been lost. For example, 4500 construction jobs lost; 2500 scientific research and permanent support personnel jobs lost; 500, 1000 visiting scientists each year lost; 5600 service and support jobs at companies doing business with the SSC lost; 14,500 indirect jobs in sales and service forever lost; at least $4 billion in construction funds that would have been pumped into Texas economy lost; at least $20 billion in direct economic benefit to the State of Texas over the life of the project lost.

So, if you are looking at the total lost to the State of Texas, we are looking at something approaching $40 billion. We haven't made a claim for that kind of money and we are not going to as long as I am chairman of this Commission or Ann Richards is Governor for those kind of dollars. All we have ever asked for, Mr. Chairman, is to be treated fairly and with the respect that we deserve as your partner.

With that, Mr. Chairman, I think I will conclude my prepared remarks and I will be happy to take whatever questions you may have.

[The prepared statement of Mr. Shelton Smith follows:]

Statement of Shelton Smith

on behalf of the

Governor's Advisory Committee -- Superconducting Super Collider

before the

Committee on Science, Space and Technology

Subcommittee on Science

U.S. House of Representatives

March 14, 1994

Mr. Chairman, Members of the Subcommittee, it is a real pleasure to appear before you today to brief you on the work of Governor Ann Richards' Advisory Committee on the Superconducting Super Collider, and to discuss the tremendous potential that exists for utilizing the valuable assets at the SSC site in a manner that will further the interests of the United States, the State of Texas, and the international scientific community.

By way of introduction, I serve as Chairman of the Texas National Research Laboratory Commission. Over the past several weeks, our Commission has worked closely with the Governor's Advisory Committee on the SSC, whose Chairman, Jess Hay, asked me to appear today on his behalf. I would like to begin by telling you a little about the Governor's Advisory Committee.

Following Congress's decision to terminate the Super Collider project, Governor Richards' immediate concern was to prevent the waste of the taxpayers' two billion dollar investment in this project by considering feasible and creative alternative uses for the tremendous assets that exist at the site -- assets that include not only state-of-the-art facilities and technologies, but also the technical expertise of the talented scientists and engineers who contributed so much to this important scientific endeavor. That, in essence, is the genesis of the advisory panel which the Governor created soon after Congress terminated the project. The charge of the panel, in short, is to make recommendations to the Governor on the future effective uses of the assets of the Superconducting Super Collider.

As the Advisory Committee fulfills its responsibilities to Governor Richards, its members also are mindful of the statutory mandate on the Department of Energy to develop a plan to "maximize the value of the investment that has been made in the project and minimizing the loss to the United States and involved states and persons, including recommendations as to the feasibility of utilizing SSC assets in whole or in part in pursuit of an international high energy physics endeavor." The Committee hopes that its recommendations will contribute in a positive and constructive way to the efforts of the Energy Department to carry out this important mandate.

Over the course of the past two months, the Advisory Committee has held extensive meetings. It has heard the detailed testimony of some 38 knowledgeable and distinguished witnesses, including Dr. Robert White of the National Academy of Sciences, who responded to the Committee's request for assistance and advice by chairing a special panel on alternative uses of the SSC assets within the NAS. The SSC Economic Development and Facilities Use Task Force received more than 60 proposals for the future use of the SSC assets from responsible sources. On February 28, 1994, the Advisory Committee submitted a report to Governor Richards outlining four key recommendations. I am pleased to have this opportunity today to share those recommendations with the Subcommittee.

The first recommendation is a fundamental -- indeed indispensable -- prerequisite to each of the subsequent recommendations I am about to detail. The Committee recommends in the strongest possible terms that immediate steps be taken to retain an essential core of at least 125 scientists and engineers, along with the necessary administrative and logistical support, through the 1995 fiscal year.

We have always recognized that the most valuable asset associated with the SSC project has been the truly superb collection of scientific talent and expertise assembled to meet the challenge of this unique project. Although the SSC project has been terminated, we still need the expertise of some subset of these scientists if we are to have the basic capability to meaningfully evaluate, assess, analyze and further develop promising proposals to maximize the use of the SSC assets in the best interests of the United States, the State of Texas, and the international community. Indeed, this core recommendation is based in large part on the January 14, 1994 report of the National Academy of Science's SSC Advisory Panel. I would like to read a brief excerpt of that report, which succinctly states the case for this recommendation:

> The immediate and critical problem is the most rapidly wasting asset -- the highly-skilled people who have been working there. They know most about the major installations such as the magnet facilities and the computer center. If significant values are to be salvaged from these facilities, enough of these people must be retained at the site to maintain the facilities and to identify, analyze and develop ideas on how they might be most beneficially used. Such people are also essential to ensure that when uses are identified the technical expertise is on site to exploit these complex facilities. The facilities alone are not enough: the expertise to understand them and to use them must be there, too.

Given the importance of this recommendation to any meaningful examination of future uses of the enormous scientific and technical assets at the SSC site, we are pleased with the recent steps the Department of Energy has taken in this regard. In a letter to Governor Richards dated March 1, 1994, Secretary O'Leary pledged that "the Department is prepared to provide financial support for scientists, engineers, and other technical personnel, as appropriate, to insure that we retain critical technical staff to help implement candidate proposals." We hope that this Subcommittee and Congress in turn will support the allocation of a modest level of funding in order to permit the retention of this core cadre of scientists and engineers.

The Advisory Committee's second recommendation to the Governor concerns the proper management of this retained cadre of scientists. The Committee recommends that the work of these scientists be managed, directed and supervised by some agency, research institution or consortium of research institutions to ensure a sharp focus on the task at hand: namely, to define, test, and recommend to the Department of Energy the highest and best uses for the SSC assets through which the value of such assets might be maximized to the benefit of the United States, the State of Texas and the international community.

The Advisory Committee's third recommendation suggests three proposed uses of the SSC assets which the Committee believes warrant further study, refinement and evaluation. The Advisory Committee found these three potential uses of the SSC assets to be among the most promising of the scores of suggestions which it has reviewed to date. These recommended proposals have been forwarded to the Department of Energy for its consideration, and I am particularly pleased to brief the Subcommittee on them today.

CENTER FOR RESEARCH IN SUPERCONDUCTIVITY

The Advisory Committee's first recommendation with regard to future uses is to assign to the retained cadre of scientists and engineers, as a matter of highest priority, the task of developing a proposal and plan to utilize the existing human resources and physical assets at the SSC Laboratory to establish a Center for Research in Uses of Superconductivity and Cryogenics.

Although some have dismissively characterized the now-terminated SSC project as a worthless "hole in the ground," in fact, the superconducting magnet facilities and associated cryogenic facility constructed at the SSC Magnet Development Laboratory and Magnet Test Laboratory are the most advanced in the world. More than $866 million has been invested in the existing facility and resources related to magnet research, development and testing. The scientists, engineers and technicians assembled to construct and operate these facilities constitute a complementary resource without parallel.

The proposed research center should focus on useful applications of the technology in such areas as high energy physics accelerator design and construction, design and fabrication of superconducting magnets, energy storage and conversion, and transportation. This proposal is fully consistent with both our national priorities for critical technologies and our national goals for enhancing the competitiveness of U.S. industry

Such a center also could constitute the core of a unique facility to manufacture magnets should the United States opt to participate in CERN, the high energy physics laboratory in

Switzerland. Using the valuable SSC facilities, this distinguished group of scientists has the ability -- unmatched anywhere in the world today -- to build the magnets that will be required by the Large Hadron Collider at CERN.

REGIONAL CENTER FOR HIGH-PERFORMANCE COMPUTING

The second concept the Advisory Committee is commending to the attention of the retained cadre of scientists is the establishment of a Regional Center for High Performance Computing. As you are all well aware, fundamental research in simulation and visualization so critical to advances in science and engineering requires expensive computational resources. The Physics Detector Simulation Facility ("PDSF") at the SSC Laboratory is already a world-class high performance computational facility, and it is in place. The PDSF is a complete parallel-computing system capable of 7,000 million instructions per second at a fraction of the cost of mainframe or supercomputers. Computer resources at the SSC Laboratory also include CAD/CAM equipment and some 3,000 Apple, DOS, and Unix work stations. To date, approximately $55 million has been invested in computer equipment at the SSC Laboratory.

The Advisory Committee believes this proposed Regional Center for High Performance Computing is a worthwhile concept given the availability of a high capacity, completed facility, the current need in the region for a better system for education and research in high-performance computing, and the existence of well qualified faculty in place to make proper use of such a center.

REGIONAL MEDICAL TECHNOLOGY CENTER

The third concept which the Advisory Committee believes merits further development is the utilization of the SSC's linear accelerator ("LINAC") and its existing physical facilities to establish a Regional Medical Technology Center to (a) conduct cancer research; (b) produce radioisotopes; and (c) provide proton therapy to patients in need to such treatment.

Specifically, the proposed center will meet a number of important national needs:

- First, the center will provide a reliable domestic supply of medical radioisotopes that U.S. radiopharmaceutical manufacturers now have to purchase outside of the country. The center also will provide experimental radioisotopes critical to research and development of clinical applications in nuclear medicine and related industrial uses.

  Medical radioisotopes are used in both cancer radiation therapy and in medical imaging of the dynamic functioning of the internal organs and organ systems of the body, which permits highly effective diagnoses and better targeted treatments.

- Second, the center will serve as the sole proton cancer therapy unit for the Central United States, and will greatly augment cancer research in the Southwest.

> Proton cancer therapy provides a means of killing cancer cells with high specificity within the body, often at sites otherwise untreatable or inaccessible to surgery, or in certain instances, avoiding an unacceptable level of damage to healthy tissues caused by other traditional therapeutic modalities such as photon radiotherapy and chemotherapy.

These initial concepts are only three of numerous ideas which eventually should be considered by the retained cadre of scientists and engineers. However, based on the Advisory Committee's extensive deliberations to date and the advice of many distinguished scientists and other experts, it has concluded that the proposed Center for Research in Superconductivity, the Regional Center for High Performance Computing, and the Regional Medical Technology and Therapy Center hold the greatest promise by which the Department of Energy might best fulfill its statutorily mandated responsibility to "maximize the value of the investment that has been made in the project".

The Advisory Committee's final recommendation urges consideration of a proposal submitted by the Nature Conservancy of Texas for the restoration of a significant portion of the 10,000 acres purchased for the SSC to its native condition as part of the nation's Blackland Prairie. The Committee believes that implementation of the proposal would provide significant research, educational, economic and environmental benefits.

In summary, I hope my testimony this morning has contributed to a fuller understanding of the tremendous potential that exists today for exploiting the valuable scientific and technical assets at the SSC site. We look forward to working with the Department of Energy in a true spirit of cooperation as it develops a plan that not only fulfills its statutory mandate to maximize the value of the substantial investment made in the project, but also the obligation we all share to the taxpayers of this country.

This concludes my prepared statement. I would be pleased to respond to any questions you may have.

Mr. MINGE. Thank you. I am impressed that in Texas things are so nonpartisan, and I will follow with interest the developments in your fine State.

I'd like to first call on Mr. Boehlert for questions. Mr. Boehlert?

Mr. BOEHLERT. Thank you very much.

And thank you, Mr. Smith, for what you are doing. I really appreciate people like you who are in the private sector who come forward and volunteer your time and talents to serve the public, and I really mean that in all sincerity.

I understand your strong feelings about the potential lost to Texas of what might have been. But I will want to point out that the American taxpayers are losing a lot too. It didn't work out the way we wanted it to from your perspective or from mine. I would also like to point out that initially I was a supporter of the Superconducting Super Collider, but it just didn't work out.

Having said that, I certainly agree with my colleague, Mr. Barton, and I could not imagine a more effective advocate than he has been, and I want to be fair. So we'll go from there.

Let me ask you this. Do you think that DOE has committed to a follow-on project in Texas?

Mr. SHELTON SMITH. Not to us.

Mr. BOEHLERT. To anybody?

Mr. SHELTON SMITH. Not that I know of. You have read the Agreement in Principle that Secretary O'Leary and I executed. Congressman, we have agreed to make our best good faith efforts to examine these proposals, to try to be faithful to what you told us to do—I mean you as the United States Congress—which is to maximize the $2 billion investment that has been made in this project.

We have agreed to a framework and a process to do that. But to the best of my knowledge, nobody has come to an agreement as to a specific project.

Mr. BOEHLERT. You know, you mention the medical project. Are you aware that the National Academy report has some real reservations about that project? I mean, I could quote specifically from the report, but are you familiar with that?

Mr. SHELTON SMITH. Yes, I am.

Mr. BOEHLERT. All right. And you also, I would hope, know that the HEPAP—the High Energy Physics Advisory Panel—members have expressed some concern about a possible new start of a project in Texas at a time when the high energy physics budget is flat and we don't have enough resources to go around for the ongoing activities at other sources. You are familiar with that?

Mr. SHELTON SMITH. Yes, I am.

Mr. BOEHLERT. All right. Well, I don't want to belabor any of the points mentioned. I think your testimony is very good testimony. You are a very effective advocate. I can only assure you, as I have tried to assure my friend and colleague Mr. Barton, that I do wish to be fair. But I am very mindful of a broad responsibility I have, not just to my constituency, but to a far larger constituency.

So, once again let me thank you very much, and let me pledge to you my commitment to be fair.

Mr. SHELTON SMITH. Thank you, sir. That's all we have ever asked for.

Mr. MINGE. The chair recognizes Ms. Johnson from Texas.

Ms. EDDIE BERNICE JOHNSON of Texas. Thank you, Mr. Chairman.

I am also grateful to the public service that, Mr. Smith, you have rendered and are going to render to the State of Texas and the people of the Nation.

I would like to hear from your vantage point what you feel the possibility might be for the magnet development to be marketable, perhaps, to CERN, to a lab outside this country.

Mr. SHELTON SMITH. Ms. Johnson, the Congress has instructed us, and when I say us I mean my Commission and the Department of Energy, to try to come up with a plan to maximize the value of these assets. We have a Magnet Development Facility and a Magnet Test Facility in Waxahachie Texas, that is the premier facility of that type on the face of this planet. There is nothing that even comes close to it.

So far we have invested approximately $866 million in this facility. It is up and running. It could be producing magnets next week if somebody told them to start producing magnets.

It would be, I believe, and Governor Richards has said this from the outset. It would be a sin to let a facility, a magnificent facility like that to be dismantled or abandoned.

Now, the U.S. Congress has to make this policy decision. I can't do that. But if it is the will of the Congress that since we are not going to build the SSC at this point that we should participate in the LHC in Geneva, then there are obviously two ways to do that. We can write a check and send it to CERN or we can attempt to negotiate an in-kind contribution.

The LHC is supposed to be started, the construction, next year. The first—the tunnel is there. The tunnel is in place and what they need first are magnets, and what better way for the United States to buy its way into LHC than by building the magnets in Texas, shipping them over as our in-kind contribution to the LHC?

I think that's something that probably makes a lot of sense and probably something we should do. But that's something that the Congress will have to determine.

Ms. EDDIE BERNICE JOHNSON of Texas. What kind of cost are we talking about to maintain this facility to produce these magnets?

Mr. SHELTON SMITH. Well, there is no capital cost, since the facility is essentially built and is running. There would be some operating cost on an annual basis. I have seen—and I am not going to tell you that I know, but I have seen various projections, from $20 to $40 million a year.

But every one, bar none, who have made those projections have told me that they thought that that was an economically viable enterprise that could produce income, generate revenue, and conceivably could become a profit center.

Ms. EDDIE BERNICE JOHNSON of Texas. So the operating costs could conceivably be an initial investment that would go on to eventually finance itself?

Mr. SHELTON SMITH. Exactly.

Ms. EDDIE BERNICE JOHNSON of Texas. And you are saying that the capability for producing and developing these magnets is currently in an operational state and all that is needed at this point is to have the authorization and funding by the Congress to continue to produce or develop the magnets?

Mr. SHELTON SMITH. Yes, ma'am. This facility is built up and running. The cost to move it, I don't think anybody really knows. But if the decision were made to dismantle the Magnet Test Facility and the Magnet Research Laboratory and the cryogenics facility which is in the cost to dismantle that and move it to California or to Brookhaven or to Fermi would probably exceed what it cost to build it.

Ms. EDDIE BERNICE JOHNSON of Texas. Thank you very much.

Mr. SHELTON SMITH. Thank you.

Mr. MINGE. The chair recognizes Mr. Barton.

Mr. BARTON. I thank the chairman.

I apologize that there are not more members of the committee here to hear your testimony because I think what you have done is in some ways the most valuable thing that's been done the last 5 months.

I want to point out for the record that Mr. Smith is double-hatted. He is chairman of the Texas National Research Laboratory Commission, and they have an obligation to the State to oversee all scientific activities that are done under the auspices of the State of Texas. He is also serving as a member of the Hayes Committee—the Hay Committee, which is a specific committee set up specifically to look at alternatives for the Super Collider. So he is actually dual-hatted. He is testifying today as chairman of the TNRLC, however.

I noticed that in the alternatives that the Hay Committee has sent to the Secretary of Energy that there is not a true high energy physics alternative. Is that because it was the committee's consideration that the dollars available would not allow for such an alternative to be pursued, or did you simply not think that any of the high energy physics alternatives submitted to the Hay Committee were worth pursuing?

Mr. SHELTON SMITH. Well, as—I think you mentioned this earlier, Congressman Barton. We have looked at, I think it's approaching 100 it's well over 70, I know—but approaching 100 different ideas for what should be done.

With all due respect, I would have to take issue with you—with your question just a little bit. I think the magnet research and development lab and the test facility and the cryogenics facility does have the potential to make a contribution to high energy physics.

Mr. BARTON. I would agree to that.

Mr. SHELTON SMITH. For CERN. As far as a pure high energy physics project standing alone at the Texas site, no.

Mr. BARTON. Okay. But we did hear Dr. Peoples indicate that there should be some review of pure high energy physics alternatives. That's the only reason I asked the question. Because my understanding is that—and you might want to elaborate on this—the marching orders that Chairman Hay has given the committee, the Advisory Committee in terms of alternatives that you can look at and dollars expended, I know—I mean the Super Collider was a multi-billion dollar, and the alternatives that have been presented are in the tens of millions of dollars in terms of capital outlay and operating costs.

Mr. SHELTON SMITH. Correct.

Mr. BARTON. So did you have an internal constraint on dollars in terms of what you really looked at as serious alternatives?

Mr. SHELTON SMITH. Well, to a certain extent, yes. I mean we have tried to keep this within the bounds of reason. The core recommendations of the Hay Committee, where all this really started, was as a result of the National Academy of Sciences SSC Advisory Panel which told us that before we do anything, the first prerequisite is to try to maintain a core cadre of the human talent——

Mr. BARTON. Right.

Mr. SHELTON SMITH. At the SSC site. We can talk about cryogenics labs and magnets all day and all night, but the thing that makes all this work are the people. And we have been very fortunate. We have had 2000 of the finest, smartest, most brilliant people in this country—in the world, working down there. And the National Academy has told us you have to keep a small group of those people in place if for no other reason to evaluate the proposals that are coming in.

So that was our first proposal. And the number that was put on that I believe was approximately $40 million for the next year to keep these people there, just to see where we go from here.

Mr. BARTON. And when I asked the Assistant Secretary, Dr. Krebs, about what the status of that was I did not get, in my opinion, a satisfactory answer. Have you had any contact, or Chairman Hay or the Governor, any contact as a result of that recommendation being forwarded to the Secretary of Energy?

Mr. SHELTON SMITH. Well, we have been, and our attorneys here in Washington have been in constant contact, on almost a daily basis. As far as a substantive yes or no response, we have not.

Mr. BARTON. And could you—Congressman Boehlert asked a question about the medical alternative that's been forwarded. It's my understanding that the alternative that your committee has forwarded is much different than the initial alternative that the National Academy of Sciences reviewed. Specifically, that the alternative that you have undertaken to support in some fashion mandates the creation of radioisotopes that can be sold here in the United States for which there is a market for which there is currently no available supply domestically. Would you elaborate on that a little bit?

Mr. SHELTON SMITH. Yes, Congressman Barton, you are exactly right. The radiopharmaceutical companies and manufacturers of this Nation have no source, no American source for radioisotopes, medical radioisotopes that we are talking about. All of that today, 100 percent is imported. We have the ability using the linac, which is not totally complete but close——

Mr. BARTON. But it's been said partially. I think for the record we should stipulate that the number, at least subject to technical clarification, it's over 90 percent complete. Is that your understanding?

Mr. SHELTON SMITH. Yes. For a small capital investment we could finish that linac.

Working with the University of Texas system and other colleges and universities in Texas, Oklahoma, Arkansas, Louisiana, it is our goal to try to set up a consortium to build a cancer proton therapy treatment center there that would serve the Midwestern United States. That would be a source of production for these radioisotopes and would serve a critical need for this Nation.

I think, as you know, right now, Congressman, there are only two such facilities like this in the country. There is one in Loma Linda, California, and there is one in the Harvard-M.I.T. complex somewhere.

There's a long waiting list for people to get into these facilities. The entire center in the South and southwestern part of this Nation doesn't even have such a facility. And we have got this ability at the Ellis County site, with a small investment and a partnership with private enterprise and universities, to build something that the people of this country need and could use.

Mr. BARTON. But I think the main point we want on the record is that the initial review, the National Academy of Sciences on the medical use was of a totally different concept than what has been actually forwarded. I mean this radioisotope idea actually has the ability to cash flow in a positive way if the capital investment is made.

And my final question is now that these initial alternatives have been forwarded and your—the TNRLC on behalf of the State of Texas has signed this initial agreement with the Department of Energy, what plans, if any, are underway today—today—to create these DOE panels to seriously review these alternatives in order to make the recommendation to this subcommittee in mid-April to mid-May?

Mr. SHELTON SMITH. I'm not sure I can answer that question, Congressman Barton. I would hope DOE is moving in an expeditious manner to do that but——

Mr. BARTON. To your knowledge then, there is nothing being done?

Mr. SHELTON SMITH. Well, other than what I have heard about Sidney Drell's panel.

Mr. BARTON. Okay. Which is different.

Mr. SHELTON SMITH. Yes.

Mr. BARTON. Now that's a High Energy Physics Advisory Panel to the Secretary of Energy which is basically a—not a ceremonial, it is much more than ceremonial, but they are a national group that meets four times a year and they have a separate task by the Department of Energy to report back in May on the status of high energy physics in the United States. That is an entirely separate matter.

I am talking about within the Department of Energy specifically tasked to work with your committee, your commission on looking at the alternatives that you have evaluated and any others that they feel may need to be evaluated, they being the Department of Energy.

I get the impression from Assistant Secretary Krebs' testimony that that's something they are getting ready to do but they have not done. And so, to your knowledge, has there been a contact directly between you or anybody on the commission and a group specifically tasked within the Department of Energy to begin to look at these alternatives.

Mr. SHELTON SMITH. No. And I don't believe they have formed their official group yet.

Mr. BARTON. Okay.

Mr. SHELTON SMITH. If they have I'm not aware of it.

Mr. BARTON. Okay. I yield back to the chairman.

Mr. MINGE. I have one brief question.

Mr. BARTON. Mr. Chairman, could I ask unanimous consent that Congressman Johnson's opening statement be put into record?

Mr. MINGE. Without objection.

Mr. BARTON. Thank you, sir.

Mr. MINGE. I have one brief question, Mr. Smith. There was some discussion earlier with the panel that preceded you about peer review, and would the Governor's Advisory Committee or the Texas National Research Laboratory Commission object to a peer review process for evaluating the proposals for use of the site?

Mr. SHELTON SMITH. No. We wouldn't object to that at all, Mr. Chairman.

Mr. MINGE. Are there any further questions?

Well, we certainly appreciate your appearance here this morning. I commend the State of Texas for what it has done, what it has invested in this project. I know that this is a deep disappointment to your State. I can think back many years ago when this project was first discussed and there were many states in this country that were vying to be the site of this tremendous facility, and that included the State that I am from, and it is with a certain irony that one now looks back and sees what has happened in that intervening period of time.

I know that this committee and subcommittee looks forward to working with the State of Texas and with your Commission on an ongoing basis to try to make sure that the site is used in a manner that benefits the people of this country to the greatest extent possible.

Thanks again for your appearance.

Mr. SHELTON SMITH. Thank you, Mr. Chairman.

Mr. MINGE. This hearing is then closed.

[Whereupon, at 12:42 p.m., the subcommittee was adjourned.]

APPENDIX

COUNCIL FOR
CITIZENS AGAINST GOVERNMENT WASTE

For Immediate Release
March 15, 1994

Contact: Leslie Paige
(202) 467-5300

WATCHDOG GROUP SLAMS NEW FUNDING FOR DEFUNCT SSC

(Washington, D.C.) -- The Council for Citizens Against Government Waste (CCAGW) today blasted a proposal by the Department of Energy (DOE) to spend $180 million in new federal money in Fiscal Year (FY) 1995 for ongoing termination activities associated with the now defunct Superconducting Super Collider (SSC). The request comes even though the DOE expects to end FY 1997 with a $40 million ***surplus*** in its SSC shutdown fund. The group also warned Energy Secretary Hazel O'Leary and Congress against the idea of trying to entice other projects to the site, which will surely lead to more pork-barrel spending.

"The American taxpayers said it loud and clear last year. Kill the SSC! Now the DOE, along with the state of Texas, is angling for another $180 million and may want to find a new use for SSC site. My question to Secretary O'Leary is 'what part of NO don't you understand?' " said CCAGW President Tom Schatz. CCAGW created and led Organizations OPposed to the Superconducting Super Collider (OOPS!), a broad-based coalition which worked successfully to have the SSC terminated in the FY 1994 budget. "Congress made reasonable provisions for shutdown costs in its FY 94 budget. They should vigorously resist the impulse to spend any new money on the SSC. It was a huge boondoggle while it was alive and now it threatens to haunt us from the grave," continued Schatz. "It's true what they say about government programs...they just never die."

CCAGW is a 600,000-member lobbying organization dedicated to seeking enactment of waste-cutting legislation.

1301 Connecticut Ave., NW
Suite 400
Washington, DC 20036
202-467-5300

FRIENDS of the earth

PRESS RELEASE

For immediate release
Tuesday, March 15, 1994

Contact: Ralph De Gennaro
202/ 879-4286

It's easier to kill a vampire!
Environmentalists urge caution on new money for SSC or its site

As one of the members last year of Organizations Opposed to the Super Collider (OOPS), Friends of the Earth was alarmed to hear that the Administration and Congress are considering spending additional money on any part of a project which Congress has already voted to kill. Especially given the extremely tight overall budget situation, any new SSC spending should have to pass the highest standards of justification.

"Last year we killed it dead. Now they say it's hungry again. I'm worried that the SSC is trying to eat our lunch," said Ralph De Gennaro, who directs the Appropriations Project at Friends of the Earth. He said that, under current budget realities, "More money for the SSC means less for everybody else."

There is no "extra" money for anything, including the SSC, in the Appropriations Subcommittee on Energy and Water Development. The Subcommittee was already sending signals that President Clinton's budget requests for environmental programs such as renewable energy research and nuclear weapons cleanup might not be fully funded, according to De Gennaro. Increasing this funding pressure are election year member requests to the Subcommittee for pork barrel water projects. Meanwhile, overall domestic discretionary spending is being squeezed by proposals to build new "walls" around the defense and crime budgets. This means the Subcommittee might not get the 602(b) allocation it needs just to fund the President's initiatives.

The idea of giving new money to the SSC does not pass the laugh test. The $180 million for FY95 would just add money for termination costs which have already been funded. Similarly, the Secretary says that she wants to spend $20 million to $40 million annually at the SSC site in future years, but again has provided no justification.

Unless there is compelling new justification for new SSC spending, Friends of the Earth believes that it's time to rally the OOPS coalition again to drive a wooden stake through the heart of the SSC and any projects at the site.

For the record, Friends of the Earth opposed the SSC on the grounds of budget tradeoffs last year, and not because of any particular harm that the SSC might bring to the environment.

* * *

1025 Vermont Avenue, NW, Suite 300 • Washington, DC 20005 • Phone: (202) 783-7400 • Fax: (202) 783-0444 • EcoNet ID: foedc@igc.apc.org
Jane Perkins, President

PROJECT ON GOVERNMENT OVERSIGHT

March 15, 1994
Contact: Daniella Brian

IS THE SSC TRYING TO RISE FROM THE ASHES?

It appears that the rumors of the demise of the Superconducting Super Collider (SSC) are greatly exaggerated.

The DOE requests for funding to terminate the project continue to balloon. As recently as January 14, 1993, then-DOE Secretary Watkins estimated the total cost of terminating the project at $278 million. At Secretary O'Leary's request however, Congress opened up the federal coffers and appropriated more than double that amount -- $640 million for this task -- without prohibiting DOE from requesting additional funding for the shut-down. At that time, DOE raised the specter that they might ask for more than $1 billion -- simply to END the project. Now DOE is making good on its threat by requesting an additional $180 million for its ongoing efforts to close down the project -- with no justification for this additional infusion. This remarkable act of bravado comes just after last month's DOE estimate that there would actually be a $40 million surplus in the termination account at the end of the process.

It certainly appears that the DOE is throwing money at Waxahachie, Texas as a political payoff rather than basing requests on any justifiable need. It unfortunately looks as though the closing of the SSC facility is being guided by politics in the worst sense of the word.

One would think that with so much money being spent to terminate the project, DOE would certainly be able to do a bang up job of closing shop forever. Yet DOE has actually issued a "Call for Expression of Interest" asking for the submission of proposals for new uses for the SSC site. This thinly veiled attempt to keep the facility running makes it difficult to understand why we need to spend so much money closing it -- unless, perhaps, the money is needed in fact to keep the lab afloat until it can be anointed a new mission.

In this extraordinary move, the DOE is circulating a request for ideas to justify keeping the facility open, rather than going through the standard practice of establishing a need for a research program BEFORE choosing its location through a nationwide search (as was originally done for the SSC program). While the SSC site is metaphorically all dressed-up with nowhere to go, it hardly seems incumbent on the DOE to take it out for a night on the town. The DOE needs to end this courtship before it wakes up in the morning with no respect.

2025 Eye Street, NW Suite 1117 Washington, DC 20006-1903 (202) 466-5539 FAX (202) 466-5596

NATIONAL
TAXPAYERS
UNION

NEWS

FOR RELEASE:
MARCH 14, 1994

FOR FURTHER INFORMATION CONTACT:
JILL LANCELOT (202) 543-1300

Statement of Jill Lancelot
Director of Congressional Affairs
National Taxpayers Union
on
The Administration's Request for Additional Funding
for the
Superconducting Super Collider (SSC)

(Washington, DC) -- The National Taxpayers Union adamantly opposes the Administration's request for additional funds to terminate the Superconducting Super Collider (SSC) and any plans to commit the federal government to a new project at the SSC site.

The request for $180 million in fiscal year 1995 is entirely unjustified given that the Department of Energy's (DOE) own termination plan for the SSC states that DOE has more than enough money on hand to close down the project. Indeed, the termination plan -- released the same day as the budget request -- shows a $40 million surplus in the termination account at the end of fiscal 1997!

We are amazed that an agency would, in this budget climate, come forward with an additional, unexplained funding request for an activity that is already overfunded. Congress terminated the SSC precisely because of this sort of cavalier attitude toward spending constraints.

In addition, we are deeply concerned that DOE appears to be committed to starting a new project at the SSC site. Taxpayers have given Congress a clear mandate to end wasteful spending projects like the SSC. They will not accept attempts to fund equally dubious programs through the back door. To start a new pork barrel project in Texas would be the ultimate misuse of taxpayer funds and would make a mockery of Congressional intent in shutting down the SSC.

The National Taxpayers Union will do everything in its power to prevent the waste of more taxpayer money on the SSC.

- 30 -

325 PENNSYLVANIA AVENUE • SOUTHEAST • WASHINGTON DC 20003 • TELEPHONE 202-543-1300 • FAX 202-546-2086

9780160445828